Comparing Apples, Oranges, and Cotton

Frank Uekötter is a Reader in Environmental Humanities at the University of Birmingham.

Frank Uekötter (ed.)

Comparing Apples, Oranges, and Cotton

Environmental Histories of the Global Plantation

Campus Verlag
Frankfurt/New York

Distribution throughout the world except Germany, Austria and Switzerland by

The University of Chicago Press
1427 East 60th Street
Chicago, IL 60637

Published with support from the VolkswagenStiftung
and the Rachel Carson Center for Environment and Society.

Bibliographic Information published by the Deutsche Nationalbibliothek.
The Deutsche Nationalbibliothek lists this publication in the Deutsche Nationalbibliografie; detailed bibliographic data are available in the Internet at http://dnb.d-nb.de
ISBN 978-3-593-50028-7

Cover design: Campus Verlag GmbH, Frankfurt-on-Main
Cover illustration: Application of pesticides on a pineapple plantation © Deutsches Museum, Munich
Typesetting: Campus Verlag GmbH, Frankfurt-on-Main
Printed on acid free paper.
Printed in the United States of America

This book is also available as an E-Book.

For further information:
www.campus.de
www.press.uchicago.edu

Contents

Rise, Fall, and Permanence. Issues in the Environmental History of the Global Plantation

Frank Uekötter

Orange juice has long emerged as a staple in the American diet. It receives almost universal acclaim for its fresh taste and its health benefits, with consumption reaching across divisions of class, race, region, and gender. Florida has dominated production ever since orange growers discovered juice as an outlet for surplus production in the early twentieth century. The state of Florida established a Department of Citrus in 1935. The industry took off after the patenting of a method to produce frozen concentrated orange juice in 1948, and corporate America entered the ring: Coca-Cola bought the Minute Maid brand in 1960 while Pepsi acquired Tropicana in 1998. The Florida legislature declared orange juice the official state beverage in 1967.[1]

It may soon be over. A disease called citrus greening is wreacking havoc to an ever growing number of orange groves all over the peninsula. Caused by a bacterium, it spread through the Asian citrus psyllid, an invasive species that was first found in Florida in 1998. Citrus greening makes trees loose foliage and causes fruit to turn bitter and drop from trees before they are ripe, effectively rendering orange trees unproductive. No known cure exists for the disease, and attempts to curb the bug's spread have met with mixed success at best. The epidemic follows on the heels of a canker epidemic that cut Florida citrus production by roughly one third. After a campaign that cost $ 600 million and included felling 12.7 million citrus trees (about ten percent of Florida's commercial acreage), the U.S. Department of Agriculture found that the fight against canker was lost and cancelled its eradication efforts in 2006. In short, a tiny insect is currently pushing a nine billion dollar industry into oblivion.[2]

---

1 Alissa Hamilton, *Squeezed. What You Don't Know About Orange Juice* (New Haven and London: Yale University Press, 2009); http://www.flheritage.com/facts/symbols/symbol.cfm?id=14 (retrieved May 27, 2013).

2 Susan E. Halbert, Keremane L. Manjunath, "Asian Citrus Psyllids (Sternorrhyncha: Psyllidae) and Greening Disease of Citrus: A Literature Review and Assessment of Risk in

The story of citrus greening mirrors the paradox of the modern plantation: the combination of permanence and notorious instability. In essence, plantation history offers a deeply ambiguous narrative—a global success story full of crushing defeats. On the one hand, plantations are a cornerstone of global food production in the modern era. They have supplied societies all over the world with a cornucopia of cheap products and will continue to do so for the foreseeable future. Western consumers can barely imagine a life without oranges, apples, coffee and other plantation products, and for good reasons: they never had to worry about them throughout their entire lives. On the other hand, plantations are constantly under threat, and many plantation systems go through cycles of boom and bust. A whole host of factors can jeopardize or terminate a plantation project, and no one knows in advance whether things will work out.

Of course, the environment was not the only source of trouble for plantation systems. Labor was a key issue, particularly since plantation economies hinged on slavery into the nineteenth century. The sugar industry in Brazil and the Caribbean, arguably the archetype of the modern plantation, is the best-known example.[3] Competition is another factor. Florida's citrus industry is not only under siege from nasty diseases but also from real estate developers and cheaper producers abroad: Brazil passed Florida as the world's leading producer of oranges some three decades ago.[4] However, environmental problems have galvanized attention long before environmentalism became a global force towards the end of the twentieth century. Soil fertility and erosion were subject to intensive debates. Pests and diseases inspired fears and frantic eradication efforts. They also inspired popular culture: the boll weevil—another tiny insect that ate its way through the Cotton South around 1900—left a mark not only on U.S. plantations but also in blues music.[5]

Florida," *Florida Entomologist* 87 (2004): 330–353; Erik Stokstad, "New Disease Endangers Florida's Already-Suffering Citrus Trees," *Science* 312 (2006): 523–524.

3 Sidney W. Mintz, *Sweetness and Power: The Place of Sugar in Modern History* (New York: Viking, 1985); Philip D. Curtin, *The Rise and Fall of the Plantation Complex. Essays in Atlantic History* (2nd edition, Cambridge: Cambridge University Press, 2002).

4 Hamilton, *Squeezed*, 213.

5 James C. Giesen, *Boll Weevil Blues. Cotton, Myth, and Power in the American South* (Chicago and London: University of Chicago Press, 2011), 95.

Scholarly Traditions

Historians have discussed the role of environmental factors in plantation history long before the rise of environmental history as a distinct scholarly field in the 1970s and 1980s. In U.S. history, the boll weevil routinely figured as the nemesis of the Old South and the main culprit for the problems of the rural South in the first half of the twentieth century, thus distracting attention from other issues such as land ownership patterns or white supremacy.[6] The Royal Botanic Gardens at Kew take pride in their role in the transfer of rubber seeds from Brazil to Southeast Asia, where plantations soon outcompeted rubber tapping in the Amazon rain forest.[7] In 1926, Avery Craven published a book with the speaking title *Soil Exhaustion as a Factor in the Agricultural History of Virginia and Maryland, 1606–1860.*[8]

It attests to the Eurocentrism of historical scholarship that these early publications did not inspire a self-conscious field of study, and this volume bears the mark of a scholarly tradition that sees Europe's role in plantation history as primarily that of a consumer. In spite of the editor's best efforts, this volume does not include an article on a plantation in Europe. That is certainly not due to a lack of suitable topics. Huge orchards produce European apples and oranges, vineyards bear the hallmarks of a plantation down to a devastating phylloxera epidemic in the late nineteenth century, and a single Bavarian region, the Hallertau, grows a quarter of the global supply of hop.[9] However, most Europeans think of plantations as an entity "somewhere else", an understanding that is perfectly in line with the word's origin. Mart Stewart's article reminds us that the plantation entered the English vocabulary with the sixteenth-century conquest of Ireland, designating what one would nowadays call settler colonization.

Looking into the environmental dimension of plantations thus follows a scholar tradition, but it is a tradition that is diverse, scattered, and widely unexplored.[10] It is also an ambiguous legacy for the discipline of environmental

6 Giesen, *Boll Weevil Blues,* xii.

7 Ray Desmond, *The History of the Royal Botanic Gardens, Kew* (2nd edition, Kew: Kew Publishing, 2007), 231–3.

8 Avery Odelle Craven, *Soil Exhaustion as a Factor in the Agricultural History of Virginia and Maryland, 1606–1860* (Urbana: University of Illinois, 1926. Reprints 1965 and 2006).

9 Christoph Pinzl, *Die Hopfenregion. Hopfenanbau in der Hallertau—Eine Kulturgeschichte* (Wolnzach: Deutsches Hopfenmuseum, 2002), 8.

10 Scholars of historiography have mostly ignored this tradition, leaving the door wide open for a study on what one might call "environmental history before environmental history".

history. On the one hand, it shows that environmental history is more than a scholarly reflection of late twentieth century sentiments: today's researchers continue a discussion that earlier scholars have long recognized as crucial. On the other hand, authors had narrowly focused on the environment as an *impediment* to plantations whereas recent scholarship views the environment more broadly as a multifaceted *context*. Furthermore, preexisting readings proved a burden as much as an encouragement. German forestry, perhaps the world's first monoculture science, produced not only coniferous plantations but also an authoritative narrative about the foresters' profession saving the country from a devastating scarcity of wood—a myth that forest historians have taken pains to dismantle.[11] In his award-winning *Mockingbird Song*, Jack Temple Kirby makes short shrift of Craven's thesis, asserting that "now, however, one must doubt that the Chesapeake country was ever lost or needed saving."[12]

Plantations are a truly global phenomenon during the modern era, but they are far from uniform. In common parlance apples and oranges make for a difficult comparison, and yet they are similar in that they are both fruit, which is not always the case for plantation commodities. Cotton is a fiber that grows around the seeds of cotton plants; rubber comes from a milky substance that *hevea brasiliensis* trees give off from incisions in their bark. Even for the same commodity, methods of production differ from region to region, and local variation exists as well; the diversity of Mother Nature dictates that there are probably no two plantations in the world that are truly identical. Scholarship has generally taken this diversity as a given. Most studies focus on a single commodity in a specific region, and many authors go to great lengths in highlighting regional and national specifics. Rarely do we find books and studies that look at more than one geographic area.

Of course, the world is a complicated place, and differentiation and academic specialization have their merits. But maybe it is time to view the plantation more comprehensively: as a global endeavor that is a key feature of modern history? Instead of leaving things at an endless variety of plantation systems, this volume proposes to see them as a transnational phenomenon that one might call the global plantation. Conditions on the ground may differ, but looking across the range of plantation systems around the globe, there are a number of things that ring familiar. In short, the global plantation

11 Cf. Joachim Radkau, *Wood. A History* (Malden, Mass.: Polity Press, 2012).

12 Jack Temple Kirby, *Mockingbird Song. Ecological Landscapes of the South* (Chapel Hill: University of North Carolina Press, 2006), 89.

is not a Weberian ideal type or an illusionary "average plantation". It is an intellectual construct that serves as a vehicle for a discussion of the common challenges for plantation systems worldwide.

It is rewarding to aim for such a problem-oriented synthesis with a discussion of environmental challenges. Unlike many other sub-disciplines, environmental history has one great potential in a global context: it holds the promise of making global history more simple.[13] Labor systems and land ownership patterns can differ endlessly around the world, but when it comes to ecological challenges, the laws of nature make for a notable degree of uniformity. Every irrigation system is coping with the threat of salinity. Pests and fungi kill plants regardless of national cultures. And when soils are exhausted from monocultures, the owner is in trouble irrespective of whether he is a plantation lord, a sharecropper, or a free peasant. (And neither does it matter if "he" is really a "she".[14]) To be sure, responses may differ depending on the socioeconomic context. For instance, planters can buy fertilizer or enlist scientific expertise that sharecroppers cannot afford. And yet the similarity of ecological challenges makes for a common thread that runs through the global history of plantations, and it seems worthwhile to explore the analytical potential of this thread. That is what this volume intends to do.

The essays in this volume are case studies on specific commodities in certain regions. But at the same time, they hold broader relevance in that they discuss issues that resonate in plantation systems all over the world. As some of these issues are discussed implicitly, this introduction seeks to identify these recurring themes and reflects on the more general implications of the case studies. It does so in a tentative fashion: the goal is to highlight perspectives for ongoing research, and to offer some ideas as to their scholarly potential. The aim is to open doors and to reflect on the challenges that students of plantations all over the world might want to explore more closely.

These perspectives are diverse and go in different directions. Here the complexity of the global plantation meets with the inherent diffuseness of environmental history. Disciplinary boundaries are never clear-cut, but those

13 Cf. Frank Uekötter, "Globalizing Environmental History—Again," in *The Future of Environmental History. Needs and Opportunities*, eds. Kimberly Coulter, Christof Mauch (RCC Perspectives 3 [2011]), 24–26.

14 Unfortunately, gender is not discussed in this collection to the extent that one would wish. Christiane Berth mentions women who came to Guatemalan plantations to join their male companions (and sometimes left after they found that it was not a good place to be), but it seems that they did not shape views about the environment to a significant extent.

of environmental history are, as John McNeill noted in a landmark article, "especially fuzzy and porous."[15] What environmental history can offer to plantation history is a multitude of hints and perspectives with varying degrees of significance and subversive power. With that, this volume is better in destabilizing established readings than in offering a new master narrative, but that arguably fits the subject. When it comes to plantations, ecological stability is as elusive as interpretative certainty. Plantations have many ways to make people 'in the know' look foolish.

While pushing conceptual and methodological limits, the following articles are also amenable to a more conventional reading. Authors were asked to write for the uninitiated and explain fundamentals of their respective plantation system, and editing sought to exorcise traces of insider code that hinders understanding beyond the circle of specialists. That makes this volume a primer for commodities in specific regions, some well-known and others less so. To be sure, the essays assembled herein make no pretense at comprehensiveness: the modern world of plantations is too wide for any such claim. But beyond their specific topics, these articles provide an idea of the richness of the overall field, and perhaps alert scholars in search of rewarding topics to a field wide open for new endeavors.

The Combination Lock: Understanding the Plantation

As we have seen, the word plantation grew out of a brutal expansionist context. It did not get much better after that. Slavery stained the image of plantations ever since the rise of the abolitionist movement, and we continue to associate the word with harsh labor, social inequality, export dependency, and monoculture. For affluent Westernites, plantations are far away, be it geographically or chronologically. Few Central European foresters appreciate it if you call their carefully managed woodlands a coniferous plantation.

From an environmental standpoint, these trepidations are revealing only in respect to prevailing mindsets. Self-definitions rarely work in a global context, and when it comes to plantations, the Eurocentric bias of popular understandings is plain. Therefore, this volume departs from what one might

15 John R. McNeill, "Observations on the Nature and Culture of Environmental History," *History and Theory* 42 (2003): 9.

call an ecological definition: plantations are first and foremost about a certain way of producing organic resources. More specifically, plantations are large, profit-driven plant production complexes that focus primarily on one commodity and cater to distant markets. With that, plantations are not necessarily about food: with rubber and cotton, this volume includes two non-food commodities. Plantations are also not necessarily about agriculture; indeed an article about commercial forests is part of this collection. In fact, forestry provides a showcase for the biological arbitrariness of Eurocentric plantation wisdom. We routinely talk about plantations when people milk latex from *hevea brasiliensis* trees, but when we talk about another tree species, and if we fell the trees rather than incise their bark, we somehow think that this is something completely different.

We will see the merits of this inclusive definition in the following, as forestry provides some of the best illustrations for the plantations' environmental intricacies. Furthermore, a broad definition frees us from the obligation to make intellectual investments in boundary work, as ways of *defining* the plantation are ultimately less rewarding in scholarly terms than ways of *understanding* the plantation. In his discussion of Liberian coffee, Stuart McCook suggests a metaphor that is worth exploring: a plantation is like a combination lock in that one false number ruins the entire enterprise. In his view,

> "the success of a commodity is determined by a set of contingent and changing problems and opportunities, each of which is like a tumbler in a lock. To open the lock, it is necessary to successfully pick a whole set of tumblers; if even one of the tumblers is not picked, the lock does not open. For any given commodity, the tumblers in this lock involve complex combinations of environmental, economic, scientific, social, cultural and other factors".[16]

According to McCook, Liberian coffee looked like a winning combination in the late nineteenth century. However, some tumblers ultimately refused to fall into line, leading to the failure of Liberian coffee as a global commodity.

It is tempting to push this metaphor further. Not all locks work perfectly: they are rusty, worn out, or in need of oiling. Similarly, plantations rarely operate smoothly, and somebody (or something) usually pays a price when the going gets tough. In fact, it is hard to say whether a plantation system is collapsing or simply facing some trouble, just as a badly damaged lock may stay in use for some time. Locks are delicate devices where repairs may cause

16 McCook, in this volume, p. 87.

performance to deteriorate, just as fixing a plantation may backfire. Plantation managers strive to achieve optimum performance in perpetuity, which is similar to owning a master key that opens every door. And then there is the mysterious attraction of locks that made them a fixture in popular culture: plantations likewise have a fascination that goes beyond the purely functional. After all, there are usually alternatives to plant production in the plantation mode, and some plantations persist in spite of the fact that their productivity is inferior to other modes of production. For example, Indonesian rubber plantations were never able to compete with smallholders throughout the entire twentieth century.[17]

Those with a theoretical bent may sense a whiff of Luhmannian sociology behind the metaphor of the lock. As one of the leading proponents of systems theory, Niklas Luhmann described modern society as a delicate set of subsystems: politics, business, academia, etc. Luhmann's key argument was that each of these subsystems was cognitively and operationally closed: all subsystems have distinct binary codes that guide their routine work. Subsystems were inherently unable to understand the code of another subsystem, ruling out targeted communication. With a view to the lock metaphor, the autonomy of different spheres rings familiar. Plantations will encounter problems in very different realms, and each of them is crucial: it is impossible to compensate for problems in the environmental realm through swift performance in the economic or political sphere, no more than one subsystem may do the job of another in Luhmannian systems theory. For modern society to work, Luhmann required all subsystems to operate smoothly, just as all tumblers in a lock need to align.[18]

In other words, plantations face environmental challenges in many different respects. Challenges may correlate, but each of them has its distinct rationale and calls for specific solutions. For example, the growth of scientific institutions may increase the problem solving capacity of a plantation system, but research may be clueless in the face of hazards such as canker and citrus greening in Florida or Panama disease for bananas. With that, there is no hierarchy in the issues that the remainder of this introduction will discuss. It completely depends on the specifics of the individual plantation

17 Michael R. Dove, *The Banana Tree at the Gate. A History of Marginal Peoples and Global Markets in Borneo* (New Haven and London: Yale University Press, 2011), 6.

18 For more information on Luhmann, see Hans-Georg Moeller, *Luhmann Explained. From Souls to Systems* (Chicago: Open Court, 2006); and Balázs Brunczel, *Disillusioning Modernity. Niklas Luhmann's Social and Political Theory* (Frankfurt: Lang, 2010).

system whether a challenge is important, critical, or irrelevant. The one thing that we can generally say is that it may be a good idea for scholars to play them through—just as people like to play with a combination lock.

Imagining the Plantation, Imagining Society

In the beginning, agriculture was simply imitation of nature. People observed the growth of grain and the movement of animals in the wild and worked from there. Plantations are too complicated for such an approach. They are complex arrangements of land and labor, climate and crops, transport and terrain that have no equivalent in nature. As a result, plantations need some kind of guiding idea, a blueprint that lays out how things should come together. Of course, this blueprint needs to allow flexible solutions on the ground, and it can have widely differing levels of refinement, but improvisation alone is an insufficient guide for management. As part of this blueprint, every plantation system produces a specific environmental imagination.

Mart Stewart's article shows the implications for the plantation societies of the U.S. South. The environmental imagination did not develop in a vacuum: It became enmeshed with ideas about social and racial hierarchies. Stewart notes that different groups have different imaginaries. There was an environmental ethos of plantation owners and managers, and another one of the slaves and the post-slavery working population. Of course, these imaginaries were closely intertwined in agricultural practice, but they were distinct intellectual universes with different goals and practices that scholars need to understand in their causes and consequences.

Stewart stresses the conceptual challenges that the merger of ideas about the environment and ideas about society implies. Reading debates among planters and slaves as antecedents of today's soil conservation initiatives and environmental justice movements robs them of context: there is no way to separate green paternalism into a racist and an ecological mindset. "The improvement of soils, the improved management of slaves, the strengthening of Southern institutions and the Southern economy—all were related and reciprocal, an organic whole, a genuine ethos", Stewart notes.[19] Slaves and

19 Stewart, in this volume, p. 39.

slave owners had environmental commitments, but it is difficult to find a language that does justice to them.

Interestingly, plantations do not need to be successful to capture the imagination. As Chris Shepherd and Andrew McWilliam show, the plantation economy of Portuguese Timor was constantly changing. Authorities tried numerous types of organization, from communal plantations and indigenous chiefs as plantation managers to penal colonies with forced prison labor, and a broad range of commodities including coffee, coconut, rubber, and tea. The one constant was the disdain for swidden agriculture and the firm belief in the inherent superiority of plantation-style production, no matter what exactly it was about. Burning forests for plantation purposes was never a problem, but not so for indigenous subsistence needs.

Ideas about society and the environment converge in the trope of "civilization", as Christiane Berth shows in her discussion of Guatemalan coffee plantations. German immigrants saw the creation of coffee plantations in the second half of the nineteenth century as a "civilizing mission" that pertained to both humans and nature. As befits Germans abroad, order was a key theme: against the chaotic backdrop of a tropical jungle, the orderly rows of plantation trees represented the ultimate victory of the human will. Of course, Western supremacy was a relative thing on the ground: Berth notes the "Tropenkoller" or tropical neurasthenia that befell Westerners in Guatemala. Revealingly, tropical neurasthenia became a respected disease, as everything else would have been tantamount to surrender to the environment. In societal as well as environmental respects, plantation narratives tend to be master narratives (with "master" being a noun as well as the customary adjective).

In short, coming to terms with the ecological imagination of the plantation is an enduring challenge, as scholars are wrestling not only with conflicting moral judgments but also with a lack of words. The one thing that we can clearly state is that environmental historians should be wary of naturalizations. Southerners sought to naturalize their plantations as the best, if not the only possible unit of production, thus giving slavery an air of inevitability. Against this background, environmental historians should probably stress plantations as eminently unnatural places of production. Jó Klanovicz makes this point most forcefully in his discussion of apple orchards in Southern Brazil. As such, the region was inhospitable to the fruit, but driven by the demands of consumers and the imperatives of import substitution, Fraiburgo became a city of apples.

Momentum and Makeshifts

Stewart's article, along with others in this volume, shows the extent to which plantations depended on the manipulation of the environment. Rice, for instance, needed well-engineered landscapes of canals and drains to maintain the right water level. It took hard work by slaves to create these infrastructures, but once in place, they implied momentum: the economics of rice cultivation hinged on whether long-term use justified the investment. Technological momentum was present on modern plantations from the very beginning, as the expensive mills at the heart of sugar plantations serve to attest. McCook mentions the investments for processing Liberian coffee, and Jeannie Whayne notes that it took enormous outlays of capital (including money from the federal government) to maintain Memphis as a cotton hub.

For students of large technological systems, momentum has been a loaded word ever since Thomas Hughes invoked it in his landmark study of electrification in Western societies. Hughes defined momentum as analogous to physics: "The systematic interaction of men, ideas, and institutions, both technical and nontechnical, led to the development of a [sociotechnical] supersystem [...] with mass movement and direction."[20] However, scholars subsequently stressed the tautology of his definition: the permanence of sociotechnical interaction leads to momentum, and momentum leads to permanence. With that, invoking momentum as an explanation is deceiving. McCook mentions that most Liberian coffee producers shunned the wet method of coffee processing in favor of a less-expensive dry method, which resulted in an inferior taste. Maybe the wet method would have delivered not just better coffee but also technological momentum. Or maybe it would have made Liberian coffee an even worse investment. In short, momentum is dangerously close to a simplistic retrospective designation of "success".

In addition to its tautological core, Hughes' concept of momentum has drawn criticism for nourishing a sense that technology drives history. Hughes was aware of this problem, positioning his concept of technological momentum "somewhere between the poles of technical determinism and social constructivism."[21] However, the momentum of plantation systems was

20 Thomas P. Hughes, *Networks of Power. Electrification in Western Society, 1880–1930* (Baltimore and London: Johns Hopkins University Press, 1983), 140.

21 Thomas P. Hughes, "Technological Momentum," in *Does Technology Drive History? The Dilemma of Technological Determinism*, eds. Merritt Roe Smith, Leo Marx (Cambridge, Mass.: MIT Press, 1994), 112.

not just a matter of technology: it was the interaction of technology, society, and the environment that made plantation systems so resilient to change. In fact, by bringing the environment more strongly into our narratives, we can grasp the true force of the imprint of plantations upon modern societies. By conditioning nature as well as society, plantations are akin to totalitarian systems—matters of life and death for entire economies and regions.

In short, the value of the concept lies mostly in the *consequences* of momentum. The combination of technological, social and environmental factors gives plantation systems an enormous potential to overcome and break resistance. We can barely account for the constant makeshifts and the enormous indifference to environmental damages and other problems without a term that highlights the inherent dynamic of a fully evolved plantation system. Plantations have a tendency to push for permanence *at all costs*, even if that implies the creation of a militarized state such as the one that Shepherd and McWilliam describe for Portuguese Timor.

If we acknowledge the momentum of plantation systems, we can better understand the hunting, fishing and gardening practices that Southern slaves maintained to feed themselves. They were improvisations under a hegemonic system, makeshifts that assured a survival that the hegemonic system did not guarantee by itself. Environmental historians have plenty of experience in highlighting hidden costs, and the articles in this book show that these costs had numerous dimensions and that these dimensions were closely linked. There was only a fine line, if any, between violence *in* nature and violence *against* nature.

The Biological Unification of the World

Diseases are perhaps the most obvious environmental challenge for plantation systems. First discovered in Ceylon and southern India in 1869, coffee leaf rust ravaged Asian plantations in the late nineteenth century and eventually emerged as a truly global plague when it crossed the Atlantic in the 1960s.[22] The Sigatoka fungus wiped out the small banana producers in Honduras while another fungus doomed Henry Ford's rubber plantation project

22 Stuart McCook, "Global Rust Belt: *Hemileia vastatrix* and the Ecological Integration of World Coffee Production since 1850," *Journal of Global History* 1 (2006): 177–195; 178.

in the Amazon rain forest.[23] The devastation brought by the boll weevil and citrus greening have already been mentioned.

The global exchange of species and pathogens is familiar terrain for environmental historians. In 1972, Alfred Crosby published *The Columbian Exchange*, a title that became the emblematic term for biological transfers across the Atlantic after 1492, and Emmanuel Le Roy Ladurie proposed his argument of a microbial unification of the world between the fourteenth and seventeenth century in 1973.[24] Since these landmark publications, a burgeoning literature has been discussing the exchange of biological material around the globe and the moral stakes involved; with disputes over "biopiracy" and "patents on life" and negotiations pursuant to the UN Convention on Biological Diversity of 1992, the worldwide transfer of plants has become a hotbed of global politics. Recent experiences with swine flue, HIV or Ebola would seem to make the case for the globalization of pathogens even stronger. However, this volume suggests a few caveats: in spite of centuries of trade and biological exchange, we still do not have a situation that one could describe as a uniform global contamination.

Contamination makes headlines, but for environmental historians, the absence of contamination is no less exciting. Berth notes the relatively benign situation in Guatemala: coffee plantations were wrestling with a number of fungal diseases, but they did not face the devastating coffee leaf rust until the 1970s. Can we provide a better explanation for these situations than sheer luck? The cognitive status of diseases is also more slippery than one would initially assume. As McCook notes, the resistance of Liberian coffee to leaf rust was ambiguous: trees fell ill, but they still produced large quantities of cherries. Liberian coffee plants were also highly variable in their disease profile. In the case of Portuguese Timor, it must remain an open question whether leaf rust was really absent or not recognized as such. Even when we accept the thesis that the microbial unification of the world will eventually take place, the effects could vary enormously. Michael Roche notes that sirex, a minor pest in Great Britain, became a huge threat in New Zea-

23 John Soluri, *Banana Cultures. Agriculture, Consumption, and Environmental Change in Honduras and the United States* (Austin: University of Texas Press, 2005); Greg Grandin, *Fordlandia. The Rise and Fall of Henry Ford's Forgotten Jungle City* (New York: Picador, 2009).

24 Alfred W. Crosby, *The Columbian Exchange. Biological and Cultural Consequences of 1492* (Westport: Greenwood Publishing, 1972); Emmanuel Le Roy Ladurie, «Un Concept: L'Unification Microbienne du Monde (XIVe–XVIIe Siècles)," *Schweizerische Zeitschrift für Geschichte* 23 (1973): 627–696.

land as it attacked the stands of *pinus radiata* that were the backbone of the country's plantation forestry. In short, the International of plant diseases is somewhat reminiscent of the Socialist International: there is an abundance of international links, but at the end of the day, the issues that count are first and foremost local.

The worldwide spread of diseases coevolved with the globalization of cures. For example, pesticides went global with amazing speed—and, as Klanovicz shows, so did the controversy over their use. And yet there was no miracle cure: keeping infections at bay was a perennial fight, one of a number of seemingly petty struggles with huge implications that characterized life on the plantation. As John Soluri remarked on banana plantations in Honduras, "Viewed from the ground level, export banana production appeared more like a series of improvisations (both creative and destructive in nature) than a well-scripted global power play."[25]

Precious Space

When it comes to plantations, the value of space is a relative thing. France famously gave up Canada in the 1763 Treaty of Paris in order to regain British-occupied Guadeloupe and Martinique: after all, the sugar plantations of the West Indies looked far more attractive from a mercantilist perspective than a cold and barren wasteland. As such, the plantation was a space-saving invention, as it maximized production on a given plot of land. However, that did not preclude a plethora of conflicts. Even in sparsely populated New Zealand, plantation forestry was competing with other types of land use.

Of course, environmental historians can claim no exclusive property to the spatial dimension of plantations, but they can add a few significant insights. Shepherd and McWilliam describe how authorities dealt with precious space through a hierarchy of resources: there were rich and poor products, and authorities urged their subjects to plant accordingly. Portuguese Timor is also a place where the Western idea of space clashed with indigenous notions: the concept of *lulik* implied that planting required spiritual consent in certain places, including places that colonial authorities saw as prime agricultural land. Scholars are also reminded that plantations are not

25 Soluri, *Banana Cultures*, 217.

strictly monocultural. Berth notes that Guatemalan coffee plantations routinely included shade trees, fields for the production of basic foodstuff, and a land reserve where cultivation would start when commodity prices rose. On São Tomé, twentieth-century cocoa plantations were also home to a variety of secondary cash crops and food for sustenance and livestock. The island is also a good showcase for the symbolic value of land ownership, as it lifted the happy owner above the strata of plantation workers.

The spatial dimension of plantation complexes was usually at odds with the spatial outreach of nation-states, and that created tensions in several dimensions. As Marina Padrão Temudo shows, the tiny sugar islands of São Tomé and Príncipe offered enough space for flourishing maroon communities until the colonial government extended its sphere of influence across the entire island. Guatemala saw a different type of conflict when Germans produced one-third of the country's coffee and even claimed a sixty percent share in some areas. The power structures of plantations coexist uneasily next to those of the nation-state: French officials in Indochina likewise expressed concern about rubber plantations being beyond their control, and scholars have called plantations on São Tomé and Príncipe "independent fiefdoms".[26] Even militarized regimes were unable to create a uniform presence. As Shepherd and McWilliam show, colonial rule in Portuguese Timor hinged on the oppressors' ability to 'divide and cultivate', that is, divide people in order to make them cultivate crops for the state.

Plantation Minds

One of the most amazing things about plantations is the ecological innocence of their makers. More precisely, ecological innocence prevailed in the early stages of plantation development and typically receded as people gathered experience. In Portuguese Timor and many other places, colonial authorities were firmly convinced that plantations were inherently superior and that they could work everywhere. In Guatemala, Germans dreaming about Rhine landscapes built coffee plantations, and their success surely owed less to their environmental sensitivity than to their fortune of having chosen a relatively hassle-free disease environment. In New Zealand, foresters im-

26 Pablo Eyzaguirre, cited by Temudo, in this volume, p. 244.

ported the European wood scarcity trope in spite of stark differences in landscapes and demand. A generation later, New Zealand authorities were aghast when they realized that they were lacking legal authority to control the importation of pathogens. As the present author has argued elsewhere, "In the beginning, and *only* in the beginning, intensive agriculture looks amazingly simple."[27]

Learning by doing was the dominant mode of intellectual advancement, only gradually supplemented by scientific expertise. Stewart shows that the South's conservation ethic (if we accept the word for once) grew in the fields at the hand of practitioners rather than in academic institutions or laboratories. This is not just a symptom of institutional underdevelopment: in spite of having one of the earliest and most efficient networks of agricultural expertise, Germany did not have a research station for hop until a fungal epidemic struck the Hallertau in the 1920s.[28] When it comes to plantations, academic freedom was an inopportune concept: science was a service provider. As Michitake Aso shows, even ecology, which holds the potential of being a "subversive" science, became a control-oriented science in the Southeast Asian context, with control serving the overarching goal of boosting production. Aso's essay also shows the highly uneven encroachment of science onto the plantation. While experts kept a close watch on pests, they played a marginal role in how to deal with weeds, where esthetics drove the planters' preferences: academic credentials ultimately yielded to the visual morality of a neat, orderly appearance. In the initial stages of plantation development, forest clearing drew heavily on indigenous swidden practices, which stands strangely disconnected from their otherwise disparaging view of indigenous farming. In Aso's reading, the plantation made ecology, not the other way round.

Going through the following papers, it is striking to note the absence of powerful professions. Even Roche uses the term cautiously in his discussion of New Zealand forestry, as he sketches a process of gradual emancipation from European and North American models. The creation of a forestry profession was a key tool for controlling careers and mindsets in Europe, but it hinged on a number of requirements that were otherwise rare in the world of plantations: state ownership of forests, a reform ethos within the administration, rigid, quasi-military discipline, the intellectual and monetary in-

27 Frank Uekötter, "The Magic of One. Reflections on the Pathologies of Monoculture," *RCC Perspectives* 2 (2011): 11.

28 Pinzl, *Die Hopfenregion*, 166.

dependence that European statehood implied, and decreasing pressure on wood resources. It is often forgotten that in spite of its penchant for "sustainability", the German forestry profession did not prevent the country from becoming a net importer of wood.[29]

In fact, the scientists in this volume are not only a diverse bunch when it comes to their academic backgrounds but also geographically disperse. Knowledge and experts crossed national boundaries with amazing ease. It is probably no surprise that New Zealand's foresters obtained their training in Europe and North America. But who would have guessed that Brazil's apples owe their existence to people from Algeria, Israel, and Japan? In Southeast Asia, British and French scientists simultaneously cooperated and competed with each other.

Hubs and Consumers

Environmental narratives tend to center on the place of plant growth, but they can also inform our thinking about events elsewhere. Some two decades ago, William Cronon showed how an ecological perspective sheds new light on the history of Chicago and the Great West [30] Whayne follows on his heels in analyzing the evolution of Memphis as a cotton hub. Its pivotal role seemed predestined at an early stage of urban development, as it quickly outgrew the competing city of Randolph, Tennessee. The truly significant contestations were local: malaria, yellow fever, city politics, always eventful in late nineteenth century U.S. cities, and a ferocious Mississippi river challenged urban development.

Interestingly, Memphis could afford to maintain rather traditional trading practices. Cotton factors made contracts directly with planters, and no market for futures trading evolved (whereas futures were common for coffee and grain since the late nineteenth century). However, the cotton hub was about more than trade with fibers. Memphis brought foodstuff back to plantations. Engineers located in the city helped planters to cope with drainage problems. Whereas Cronon's discussion of Chicago focused on the com-

29 Bernd-Stefan Grewe, "Das Ende der Nachhaltigkeit? Wald und Industrialisierung im 19. Jahrhundert," *Archiv für Sozialgeschichte* 43 (2003): 61–79.

30 William Cronon, *Nature's Metropolis. Chicago and the Great West* (New York and London: Norton, 1991).

modification of nature, Whayne shows that Memphis was really a comprehensive service provider for the cotton hinterland. It would be worthwhile to look at other urban centers in plantation regions and compare their significance. Does every plantation commodity need a hub? Or is the need related to a certain volume of trade or a specific level of technological development?

If we follow the commodity chain beyond the hub, we ultimately end up with the consumer. We can see the mystery of consumer preferences nicely in McCook's discussion of Liberian coffee, where taste was a key issue. That makes for several levels of complication. Taste is personal rather than collective, it is tied to cultural preferences, and it is hard to pin down for a historian. From an environmental perspective, perhaps the most interesting issue is the inconsistency of taste: biological diversity has ecological merits, but there is only so much variation that modern consumers can tolerate. It is hard to tell in retrospect whether Liberian coffee was doomed because of its weird taste or because of the absence of a "right" taste. Even more, it is hard to say how historians can arrive at an informed conclusion on this matter.

When it comes to the second half of the twentieth century, thinking about plantations, consumers and the environment inevitably brings up the issue of pesticides, along with the difficulties of making sense of them. As Klanovicz shows, the problem was not just about material dangers but also about producers and consumers living in different worlds. Late twentieth century urbanites rarely understand the sense of spraying: for them, it is simply indiscriminate use of toxic substances. Agriculturalists may find that a rather uninformed view of the matter, but then, things look a bit different when you regard them from the other end of the commodity chain.

Plantations Forever?

Most of the plantation systems discussed in this volume are still around. Guatemala and East Timor are still deeply invested into coffee, the U.S. South continues to produce huge quantities of cotton, and to Brazilians, Fraiburgo still means first and foremost apples. Of course, modern technology has changed production methods profoundly, but that arguably makes the resilience of plantation systems even more impressive. Somehow people find it difficult to abandon plantations: as this volume was going to press, the latest news from Florida is that Coca-Cola will spend $ 2 billion to plant

25,000 acres of oranges, notwithstanding the fact that, as it stands, these trees will be easy prey for citrus greening.[31]

Of course, scholarly bifurcation may be at play here: plantations that persist claim the historians' attention more easily than plantations that collapsed. However, failures can be quite revealing, and the articles of McCook and Temudo demonstrate two very different paths towards the end. In the case of Liberian coffee, a plantation system collapsed in the making, and failure helped to blaze the path for the more successful robusta variety. On São Tomé and Príncipe, the time span is vastly greater: plantation systems for sugar, coffee and cocoa left their mark on the islands over the centuries, including a Soviet-inspired period with state-run plantations where managers came from the old elite. However, the combination of fluctuating cocoa prices, famine, mismanagement and the end of Soviet power resulted in a push to diversify and change land ownership patterns—though land reform was ultimately more successful in jeopardizing traditional plantations than in ending social discrimination.

All in all, plantations offer plenty of fodder (if the metaphor is allowed) for the open-minded historian. But at the same time, a plethora of perspectives tends to create a yearning for a grand theory that brings it all together. Several papers juggle with Jim Scott's idea of "high modernism", but that concept remains strangely unspecific as to time, place, and thematic context.[32] Maybe wrestling with complexity *is* the sales pitch? One of the key insights of plantation history is that you never know where problems will come from, and environmental historians are well poised to highlight some of the more inconspicuous causes of trouble. As actors go, insects, microbes and weeds make for less-than-perfect heroes, but they deserve their place in our narratives of the global plantation—and not just because you probably care about that daily glass of orange juice.

31 *New York Times*, May 10, 2013, A14.

32 James C. Scott, *Seeing Like a State. How Certain Schemes to Improve the Human Condition Have Failed* (New Haven: Yale University Press, 1998).

Plantations, Agroecology, Environmental Thought, and the American South

Mart A. Stewart

Scholars of all kinds have studied plantations or what Philip Curtin calls "the plantation complex," mainly in the Americas. Edgar Tristram Thompson's *The Plantation: An International Bibliography*, published in 1983, lists over 1200 sources, and a bloom of scholarship about plantation slavery has added to this literature since then—at least indirectly. Much of this scholarship has been devoted to the study of plantations in the U.S. South. Plantations and plantation slavery in the seventeenth, eighteenth, and nineteenth century U.S. South have been studied in every conceivable way by historians of the region: as economic and political units, as the nexus of antebellum race relations, as public health problems, as bastions of Southern cultural distinctiveness, as expressions of planter paranoia, as agricultural units that linked household economies with global capitalist ones, as the sites of the extended families unique to the region, and as landscapes of legend and myth. All acknowledge the important agricultural function of plantations—in the production of the commodity crops, especially cotton, that fueled the southern antebellum agricultural economy. And all agree, either implicitly or explicitly, of an essential relationship between humans and nature which were arranged by plantations and those who created them. Some commentators on the subject have claimed that farms that were devoted to cash-crop agriculture, large agricultural operations devoted to monoculture that are mechanized, or modern agricultural operations that impose a strict agricultural regimen on both land and hands—the massive tomato-producing farms of Immokalee adjacent the Everglades, where a third of U.S. tomatoes are grown on highly manipulated sandy soils by poorly paid immigrant labor, for example—also should be considered plantations. But in general the scholarship on plantations in the U.S. South have focused on the seventeenth century through the 1930s, when Southern agriculture in general and at last began to modernize. They have also at least indirectly agreed with what Edgar Thompson said about plantations, that they were devoted to cash crop production, but were also an important "race-making" institution in the American South.[1]

Historians have differed, though, in their explanations of the importance of the physical environment in the development of plantations in the South and in the enduring significance of plantations as the agricultural unit in which most commodity agricultural production was conducted in the region. Plantations were not inherently more productive than small farms in southern cotton-growing regions, some historians have argued, and slavery was only one of several possible labor systems planters could have chosen to grow staple crops. Once they had created the system, the master class had to consolidate their political power to assure the continued existence of plantation agriculture and the peculiar institution, and this process, not environment and crops, spawned the replication and development of plantation agriculture. Other historians have claimed that institutional development and political class-aggrandizement were important components of the development of the cotton South, but ecological conditions were more fundamental to this development. With an argument that fine-tunes one made in a pioneering work in the 1920s by U.B. Phillips, *Life and Labor in the Old South,* these historians claim that the soil and climate (especially the relatively long growing season) of the region favored the cultivation of valu-

---

1 Curtin's scholarship on plantations argues a very large tent; he says that they originated when Europeans encountered sugar-growing in the eastern Mediterranean: Philip D. Curtin, *The Rise and Fall of the Plantation Complex: Essays in Atlantic History* (Cambridge: Cambridge University Press, 1990). Edgar T. Thompson, *The Plantation: An International Bibliography* (Boston: G.K. Hall & Co., 1983). Thompson's insight about race-making appeared in "The Plantation as a Race-Making Situation," in *Sociology: A Text with Adapted Readings*, eds. L. Boom and P. Selznick (New York: Harper and Row, 1955), 506–7. This analytical insight was ahead of its time. It emerged again in the 1980s, most prominently in a classic article by historian Barbara Fields in 1982, and has been at the core of a generation of scholarship on plantation slavery: Barbara J. Fields, "Ideology and Race in American History," in *Region, Race, and Reconstruction: Essays in Honor of C. Vann Woodward*, eds. J.M. Kousser and J.M. McPherson (New York: Oxford University Press, 1982), 143–78. Sociologist Edgar Thompson was ahead of his time in several ways—but his analytically prodigious lifetime of scholarship on the plantation has been known to historians but usually not studied by them; for an appreciation of his work and legacy, see Sidney Mintz's introduction to the 2010 publication of Thompson's hitherto unpublished 1932 doctoral dissertation: in Edgar Tristram Thompson, *The Plantation,* eds. Sidney W. Mintz and George Baca (Columbia: University of South Carolina Press, 2010), i–xx. For Immokolee tomato plantations, see Mark Bittman, "The True Cost of Tomatoes," *New York Times*, June 14, 2011, http://opiniator.blogs.nytimes.com/2011/06/14/the-true-cost-of-tomatoes/?_r=0, accessed June 18, 2011.

able staple crops, which were best and most efficiently grown on plantations using slave labor, which in turn created the southern agrarian class structure and an "enclave" economy. They do not disinter geographical determinism to explain cultural traits of Southerners, but do, indeed, look to geography and to the environment as a prominent cause of plantation hegemony and of Southern economic and political distinctiveness. Historians who look at the relationship between plantations and the environment agree that this relationship was crucial, in other words, but disagree on how this relationship worked itself out.[2]

Along with their enduring importance as sites of agricultural production, plantations in the U.S. South were the site of environmental degradation and improvement, both—which had a legacy that continued well into the twentieth century and that became part of a national problem by the second quarter of the twentieth century. The growing of staple crops year after year on the same land produced the problem of "soil exhaustion"—an environmental problem only in places where agriculture is practiced—and perhaps the most important environmental problem in the history of this region. Soil exhaustion became such a problem in the older parts of the South (in Virginia and South Carolina, especially) that many planters by the 1820s believed it would lead to the decline of the region as a whole. Some of them began to do something about it, and though many did not practice what they preached, they created a formidable body of ecological knowledge and ecological advocacy by way of their attempts at agricultural reform. In tandem with ideas about how to manage slaves more humanely, the agricultural reform movement produced an orientation to arrangements of humans and nature that constituted a model for environmental reform as well as an agroecological ethic—and a cautionary tale about agroecological ethics as well. Improving planters in general continued to treat both slaves and soils with condescension, and it needs to be emphasized that this reform movement masked no small amount of misery, but planters also began to understand

2 Ulrich Bonnell Phillips, *Life and Labor in the Old South* (Columbia: University of South Carolina Press, 2007; repr. Of Little, Brown, and Co. 1929 edition). For a critique of Phillips' moderate geographical determinism, which also indirectly argued for a white supremacist racial order to which Phillips himself subscribed, see Mart A. Stewart, "'Let Us Begin with the Weather?': Climate, Race, and Cultural Distinctiveness in the American South," in: *Nature and Society in Historical Context*, eds. Mikuláš Teich and Roy Porter (Cambridge: Cambridge University Press, 1997), 240–56.

ecological limits—and how they understood these is an important story in the larger environmental history of plantations.[3]

African American slaves—those who did most of the real work on Southern plantations—had other ideas and acted in other directions. First of all, the plans of improving planters for restoring the ecological integrity of their soils often succeeded or foundered on the willingness of their slaves to carry them out. Also, in the quarters and along the edges of fields, in their gardens and in patches out in the swamps and forests, they practiced their own agricultural and environmental ethic—to produce and reproduce family and community rather than profit and power. Their environmental practices at times reinforced plantation power relations, at times subverted them. Altogether they created the beginnings of an African American environmental ethic that reached full expression in the environmental justice movement in the region in the late 20th century (not coincidentally, most prominently in the old plantation districts of the South).

America was generally a rural nation with most Americans engaged in the work of agriculture until the early twentieth century. But the imprint of agriculture was deeper in the South, lasted longer, and almost from the beginning (at least after Europeans arrived) was driven by a set of relationships that gave landowners control over both land and labor—and a small elite of landowners control over more land and more labor. However modern scholars agonize over the meaning of the term, "plantations" in the U.S. South were at first what they had been when the word "plantation" first entered the English language by way of the English conquest and colonization of Ireland—first of all, simply a place and a colony—a place planted by the conquering culture. But plantations as conquered places where capital and labor were organized and applied to produce agricultural commodities emerged very soon after the first Southern British colonies were founded in

3 Avery Odelle Craven's study of soil exhaustion created this productive scholarly chestnut: *Soil Exhaustion as a Factor in the Agricultural History of Virginia and Maryland, 1606–1860* (Columbia, S.C.: University of South Carolina Press, 2007). The discussion was revived by Eugene D. Genovese in *The Political Economy of Slavery: Studies in the Economy and Society of the Slave South* (Cincinnati: Wesleyan University Press, 1988) and has been developed at length in Jack Temple Kirby's introduction to his examination of reform planter Edmund Ruffin's writings on marling as a solution to soil exhaustion and its attendant social and economic problems: *Nature's Manager: Writings on Landscape and Reform* (Athens: University of Georgia Press, 2000).

the Chesapeake and the Carolina lowcountry in the seventeenth century.[4] The plantation itself was an adaptation to the difficult environment that the first colonists encountered—but also one that allowed them to transfer to North America a form of agricultural production that had worked in kindred climates and soils in the Caribbean. Long growing seasons and ample moisture, and a good river system for transporting cash crops made commodity crop production possible at the same time that poor soils and the conditions of slavery forced mobility in both land and labor. By the nineteenth century, plantations were the backbone of southern agriculture and drove the economy of the region. Cotton agriculture moved from Georgia and South Carolina to Texas; the significant frontier in southern history is the cotton frontier—and slavery moved along with it. How the South as a region—given its geographical diversity and that a large percentage of landowners did not own slaves—can be identified has been an issue of perennial debate for historians of the South. But that the planter class—those who possessed plantations and slaves to work them—held most of the wealth and the power in the region and that Southern society was from the beginning at least bi-racial is beyond question.[5]

Plantation Agriculture in the U.S. South

All of the main plantation commodity crops were demanding of both labor and soil. Rice and sugar had the most demanding regimens. Rice plantations along the South Carolina and Georgia coasts yielded over 90 percent of American rice on the eve of the Civil War. Sugar plantations along the lower Mississippi (the only part of the South where the growing season was long enough for sugar to be grown with consistent success) made sugar planters

4 Jack Temple Kirby, *Mockingbird Song: Ecological Landscapes of the South* (Chapel Hill: University of North Carolina Press, 2006), 76.

5 Even though commentators exaggerate the homogeneity of the South as a way to identify it, investigators have found a high level of consensus among black and white residents of "the South" that their region has a cultural integrity. At the same time, as Edward L. Ayers concludes in a thoughtful essay on the problem of Southern regional identity, "The South is continually coming into being, continually being remade, continually struggling with its pasts." See Edward L. Ayers, Patricia Nelson Limerick, Stephen Nissenbaum and Peter S. Onuf, *All Over the Map: Rethinking American Regions* (Baltimore: The Johns Hopkins University Press, 1996), 82.

among the wealthiest of southern planters. But both crops required massive manipulations of the environment. Rice was grown in well-engineered landscapes of canals and drains that tapped the rise and fall of tidewater rivers to flood and drain the fields. These had to be dug out of freshwater swamps by hand by slaves. Sugar required two heavy investments of labor—when the rattoon was planted to make new crops in the spring, and when the relatively heavy cane had to be harvested and processed in the fall and early winter. Sugar processing required large quantities of firewood, and this also was extracted by sugar plantation slaves from forests adjacent the plantations. Cotton was a more resilient crop, did not have the particular environmental requirements of rice and sugar and could be grown wherever the growing season was long enough. It could also be grown by one set of hands as well as by several (though not in the same quantity), and was non-perishable and could be easily stored and shipped. But the difficulty of processing cotton by hand made it a crop of such marginal profitability that some southerners believed plantation agriculture would never permanently take hold outside of the original plantation districts of the colonial south.

Even non-Americanists know about the difference an efficient cotton gin made—Eli Whitney's invention in 1793 and a rapidly expanding market for cotton among British and, later, American textile manufacturers, made short-strand cotton the most important staple crop in the South and assured the expansion of the plantation system. Although slavery was not essential to the production of cotton (some of the South's cotton throughout the period was grown by the three-quarters of southern whites who did not own slaves), white southerners assumed that slavery was essential to the efficient production of large quantities of cotton. The invention of the cotton gin and the expansion of cotton culture also assured the expansion of slavery.

The spread of row-crop monoculture etched the plantation landscape into a larger proportion of the region, and the "Cotton South" took shape. By the 1820s and 1830s production west of the Appalachians, on the rich prairie lands—the "black belt"—of central Alabama and Mississippi, on rich alluvial river bottoms, and on uplands throughout the region, had also become significant and these areas became "Southern" in character. Southern rivers, especially the Mississippi, were essential to this growth, linking river "cotton ports" like Memphis to important export ports on the coast. By 1860, three-fourths of the area's cotton was moving on the South's river system through Mobile, New Orleans, and the young Texas ports.

Production of cotton in the South increased from about two million pounds in 1791 to over a billion pounds in 1860. By 1840, the plantations of the American South were producing over sixty percent of the world's cotton and throughout the period exports of cotton made up over half the earnings of domestic exports from the United States. Capital generated by cotton earnings not only supported a large class of cotton producers in the South, but attracted foreign investment and contributed to the industrial growth of the North. It's difficult to make the kind of environmental argument about cotton that anthropologist Marvin Harris makes about water regimes and that can be applied to the temporary persistence of rice agriculture on the Georgia and South Carolina coasts—that after they are created, water regimes that require extensive and expensive infrastructures have an inertia that often overcomes challenges or at least limits alternatives to those regimes.[6] But cotton agriculture had a persistence after the war that, economic and political factors aside, gained some of its momentum from King Cotton before the Civil War. Sharecroppers replaced slaves as southerners returned to a commitment to cotton that lasted until southern agriculture modernized in the 1930s and 1940s. But soils in the old cotton plantation districts of the South continue to show the long history of depletion and erosion, and the residents of these districts—many of them descendents of slaves and sharecroppers—continue to be among the poorest of the poor.

Planter Landscapes and Green Paternalism

The beginnings of conservation in the South—until recently largely ignored by American environmental historians, who have favored the kinds of environmentalism created by liberal moments and leaders who have focused on more open places in America, usually in the West—was also thoroughly enmeshed in the culture of slavery and the agricultural history of the South—and emerged by way of the experience of planters with plantations in tandem with cosmopolitan circuits of knowledge. In the 1820s and 1830s, when many Southerners began to feel their region was in the grip of a cultural and economic crisis because of soil exhaustion and outmigration but more importantly from attacks against slavery from outside the region, they sought

6 See Harris, *Cannibals and Kings: Origins of Culture* (New York: Vintage, 1991).

solutions through agricultural reform and conservation practices. Leading planters, especially in the older parts of the South where soil exhaustion most profoundly challenged the continued vitality of plantation society, advocated more beneficent management practices for both slaves and soils. They exchanged and promoted ideas about better ways to grow rice, cotton, sugar, tobacco, and other crops, about crop rotation and fertilizing, about machines that would make agricultural practices more efficient, all with the goal of diversifying southern agriculture and making it more efficient and restoring depleted lands. An ethic of stewardship emerged in the abundant discussions of agricultural improvement that showed up in addresses to agricultural societies and in the pages of new agricultural journals. Concerns about environmental problems, reappraisals of the distinctiveness of Southern culture and the Southern economy, and participation in cosmopolitan circuits of knowledge all converged in this ethic. As is the case with every set of environmental ideas, solutions to environmental problems were rooted in ideas about society; ideas about how to improve agricultural practices were driven by ideas about how to reform Southern society.

Some of the practices advocated by planters followed what modern environmentalists might recognize as ecological principles. One of the Georgia lowcountry's most progressive planters, James Hamilton Couper, for example, sought not only the salvation of tired soils, but agricultural practices that would harmonize with "the principles of vegetation." In a contribution to the *Southern Agriculturist* in 1833, he described an elaborate soil and crop management program that went further than the usual laments about soil exhaustion in its recognition of the basic unsoundness of monoculture and the implications for plantation agriculture. "Where nature is allowed to sow her own seeds and reap her own harvest," he wrote, "the earth, instead of being impoverished by her vegetable productions, seems at each new effort but to augment that fertility, which is ever presenting to the eye a varied aspect of beauty and fruitfulness." When the earth is instead controlled by humans for specific agricultural productions, though, the effects have been markedly different: "Their exhaustion generally follows production, and utter impoverishment would succeed to teeming fertility, were not resort made to benign nature, or to expensive manures, to restore the lost fertility." Once soil was used for agriculture, planters should carefully follow crop rotation schemes that "harmonized" with nature, Couper explained, if they wished to ensure perennial fertility. In the second part of the article he laid out such a scheme, one that he had worked out on the highlands and tidal swamplands of his Georgia plantation.[7]

By all indications, Couper planted what he preached, with one sixteen-year sequence of crops on the highlands of his sea-island plantation and another nine-year one on the tidal swamplands. If the records he meticulously kept are accurate indications of his planting practices, he followed these detailed rotation schemes for at least the first thirteen years of his tenure, and probably longer. Though Hopeton was never an exceedingly profitable plantation, it operated in the black while Couper was managing it. By all accounts, Couper was a model progressive agriculturist, who also organized every aspect of the plantation operations, maintained an organized system of supervision, and kept precise records of every aspect of these regimens.[8]

James Hamilton Couper represented the best of rice and sea island cotton planters, and can be considered a conservationist with a successful track record—both in the efforts he extended to preserve or restore fertility on his own lands and in his efforts to proselytize about them in the Charleston-based *Southern Agriculturist* and other venues. Improving planters in tobacco or cotton districts had a tougher row to hoe. The best example of this group, Edmund Ruffin, has also become, for environmental historians, the Ur-example of the conservation movement among antebellum Southern planters. Ruffin was to the manor born in Virginia, on a plantation, Coggin's Point, in an old tobacco and cotton farming area of Virginia—he inherited 1500 acres worn out by production of cotton and corn—and fifty slaves. He early acquired an interest in improvement, sparked more by his discovery of the literature of improvement in 18th century British agronomy, and then by Sir Humphrey Davy's *Elements of Agricultural Chemistry* (1812). Historians of science have pronounced Davy's soil chemistry wrong, but his observation of the process by which quicklime obviated the acidity of sterile soil and made it productive pushed Ruffin toward what became a mission for him—an advocacy of the application to worn-out soils of "calcareous manures," and especially of the marl, the fossilized shells, that could be found in abundance

7 James Hamilton Couper, "Essay on Rotation of Crops," *Southern Agriculturist* 6 (Feb.–March 1833): 57–66, 113–20. Couper recognized intuitively the significance of beneficial soil properties and food webs of micro- and other organisms that develop in soils that are not channeled into monocrop production. For a full discussion of Couper, planter conservation ideas, and "green paternalism" on the rice coast, see Mart A. Stewart, *"What Nature Suffers to Groe": Life, Labor, and Landscape on the Georgia Coast, 1680–1920* (Athens: University of Georgia Press, 1996), 182–88, 323–245, n74.

8 Bagwell, "Couper," 201–68; Clifton, "Hopeton", 446–49.

in his plantation neighborhood. These would lower the soil pH, and fix nitrogen in soils that had been also manured with dung and green manure, and make it available to plants instead of evaporating into the atmosphere. Ruffin's faith in marl was bolstered by exhaustive experiments that he conducted on his own plantation—and in the studies he did of soils elsewhere in the South. He proselytized relentlessly on behalf of marling and other improvements—most notably, the reclamation of swampland and the enclosure of the open range—in the agriculture journal he founded in 1833, *Farmer's Register*. Adding both green manure and marl, as well as other agricultural improvements, would save the soils and the plantations of the "exhausted" parts of the South, he argued, staunch the flow of migrants to fresh lands in Alabama, Arkansas, and Texas, and contribute to the economic strength that the South would need to fend off the intensifying attacks on its peculiar institution and distinctive way of life.[9]

Many planters merely talked about reform and did not dirty their hands with the attentive management and hard work that was required to carry reform ideas into practice. And even those who tried it often gave up before they could accomplish much—the process by which the Ruffin plan transformed soil was both slow and arduous, and many planters did not stay the course. But even armchair agricultural reform constituted an early source of conservation ideas that has only recently been examined by scholars who have studied the history of conservation and environmental ideas and politics.[10] These conservation ideas, though they had much in common with a larger movement among reform farmers throughout the older regions of the U.S. in antebellum America to stay and improve rather than skim and move, took shape in a distinctive form within the context of slavery and regional consciousness. This, too, is part of the deeply social content of Southern en-

9 Ruffin has been the subject of several biographies and scholarly discussions of his extensive writings on agriculture and Southern society. Environmental historians have also given him quite a bit of attention—most significantly, Jack Temple Kirby, in *Nature's Management.*

10 Whether we need additional scholarship on this topic is partly a matter of political predilections. The study of the thought of slaveholding Southern conservatives in any way is not a path to career success for young scholars—the best recent work on them continues to be done by established scholars with commitments to debates of a vintage that tends to be distracting to studies of the environmental thought of Southern planters. Environmental historians, on the other hand, who have studied this thought have not adequately retained the context in which these ideas emerged. As Eugene Genovese once incisively observed, attention to "who was riding whom" always needs to be paid, in any study of the antebellum South—and especially of plantation slavery.

vironmental history and more specifically the environmental history of the plantation.[11]

It is important to underline this last point, that conservation movement of antebellum reform planters took place in a particular historical context and was always connected to the social and political agendas of these planters. Ruffin and Couper and other improving planters who read broadly in the agronomic literature—who were cosmopolitan in their connection to larger circuits of knowledge—were focused on fixing things in their own locale, first of all. But they also sought to transform and strengthen the South as a region—the improvement of soils was just one of the ways they sought to do this. Another was by reforming the management of slaves. The scholarship is extensive on the development of new ideas about slaves, the acknowledgement that they had souls and were deserving of humane treatment, the advocacy of better methods for managing slaves and the employment of all of these ideas to argue that slavery was a positive good. Slaveowners saw the handwriting on the wall and attempted to improve the institution to defend it against growing attacks by anti-slavery critics. Their efforts were, as Eugene Genovese famously observed, the first time in the history that a

11 For another discussion of the emergence of ecological sensibilities among southern planters, and especially, once again, Edmund Ruffin, see Joan E. Cashin, "Landscape and Memory in Antebellum Virginia," *Virginia Magazine of History and Biography* 102 (Oct. 1994): 477–500. Richard Grove has recognized the antecedents to conservation thought among slave-holding planters, though his work has focused not on the nineteenth-century South, but on seventeenth- and eighteenth-century colonial specialists in Dutch, French, and English colonies—island colonies, especially. See Richard Grove, *Green Imperialism: Colonial Expansion, Tropical Island Edens and the Origins of Environmentalism, 1600–1860* (Cambridge: Cambridge University Press, 1995), esp. 2–15. Steven Stoll has more deliberately than any other scholar looked at the ecological sensibilities in agricultural reform in the South during this period, and links it up to similar developments north of the Mason-Dixon Line at the same time: Steven Stoll, *Larding the Lean Earth: Soil and Society in Nineteenth-Century America* (New York: Hill and Wang, 2002). But he does not adequately identify the extent to which southern agricultural reform was shaped by the institution of slavery and its place in a larger effort to reform slavery to address attacks from outside the region—and that all of the *work* of applied conservation in the South was done by slaves. The study of plantations as agroecological units and the complexity of the management strategies over time of some Southern planters would be well served by more studies of individual plantations: See, for example, chapters 3 and 4 of Stewart, *"What Nature Suffers to Groe."* Two recent works by environmental historians demonstrate how much can be learned from the study of individual plantations: Lynn A. Nelson, *Pharsalia: An Environmental Biography of a Southern Plantation, 1780–1880* (Athens: University of Georgia, 2007); and Drew A. Swanson, *Remaking Wormsloe Plantation: The Environmental History of a Lowcountry Landscape* (Athens: University of Georgia Press, 2012).

counter-revolution preceded a revolution.[12] Genovese also famously argued that planters, in their efforts to reinforce what they believed to be the organic relationships of the extended "household" family on the plantation, replaced an older patriarchal ideal with a paternalistic one—that they explained the relationship of master to slaves as one akin to the reciprocal one between father and children. Paternalism was still a matter of one class riding another and still racist, and plantation masters were still very much in the saddle. But an advocacy of those behaviors that added up to paternalism was designed to reform the management of slaves—to treat them humanely and to tend to their souls.[13] Other historians have designated Genovese's interpretation of slaveholder ideology as "quaint," and claimed that slaveholders were just like other entrepreneurs in the capitalist West, in it for the money. Plantations, says James Oakes, Charles Sellers, and others, were capitalist enterprises, and slaves were merely severely bonded laborers—distinguished from the wage-earners who worked the crops of capitalist agriculture elsewhere only by the institution of slavery. And that slaveowners, who they agree held to the family as the fundamental social institution, were merely participating in a larger emerging bourgeoisie cult of domesticity.

This is now a tired argument—and in any case, fresh scholarship on slaveholder ideology—by Eugene Genovese and Elizabeth Fox-Genovese as well as by several young scholars, have discarded it as a somewhat irrelevant one. Jeffrey Young's revisionist take on this argument is especially convincing: that slaveholders were indeed facing substantial challenges in the 1820s and 1830s, and some of these were by way of the plantations that provided their bread and butter, and that they responded to these through what he calls a 'corporate individualism" that had been evolving as a system of thought and behavior in the South for some time. Slaveholders did not give into the radical egalitarianism and individualism that was common outside the region, though they also did not reject either egalitarianism or individualism, he argues. But they continued to consider both within a framework of individuals within a corporate entity, within an organic hierarchical community, and

12 In Eugene D. Genovese, *Roll, Jordan, Roll: The World the Slaves Made* (New York: Vintage, 1972).

13 Genovese worked this out in several of his works, but most thoroughly in *Roll, Jordan, Roll.* Eugene Genovese and Elizabeth Fox-Genovese explore the intellectual and religious underpinnings to paternalism in the massive *The Mind of the Master Class: History and Faith in the Southern Slaveholder's Worldview* (Cambridge: Cambridge University Press, 2005). That slaves continued to be treated as slaves—and that paternalism masked enormous suffering and misery—needs to be noted and underlined.

remained proactively distinctive in their defense of the plantations, slavery, and the South by the way they balanced these—with the corporate entity to which individuals belonged of ascendant value.[14] The intense advocacy of conservation ideas by Edmund Ruffin and other improving planters should be seen in this context. The improvement of soils, the improved management of slaves, the strengthening of Southern institutions and the Southern economy—all were related and reciprocal, an organic whole, a genuine ethos.

Planters went further than this with ideas about the relationship between nature, agriculture, and culture. They also used nature—and a particular kind of nature, as it was arranged on antebellum plantations—as part of an argument to justify agriculture and slavery and to defend what they believed to be distinctive about southern society. In the mid-nineteenth century, the South was not just a region or a section, but also a nation—that leading southerners justified partly by a defense from nature. In the 1850s, some influential southerners developed a pro-slavery argument that *naturalized* staple crop production, slavery, and southern society. The argument went like this: Because of the climate, staple crop agriculture was the best adapted to the region (and as the defensive fever of the 1850s intensified, Southerners ignored variations in climate within what became the solid South); because of this agriculture, the plantation was the best unit of organization for growing staple crops; because of plantations, slavery was the best labor system, because Africans had been imported as plantation laborers and, according to prominent variants of the argument, were better suited for labor in the long, hot summers; because of all three, the South possessed an economic and cultural uniqueness. Pro-slavery ideologues more often defended the peculiar institution and the culture that depended upon it in arguments derived from Scripture rather than nature, but by the end of the 1850s, "the sunny South" and the "peculiar climate" had become a fundamental point in an ideological defense, a note in a common chord struck to reinforce the commitment of leading Southerners to slavery and to Southern society.[15]

---

14 Jeffrey Young, *Domesticating Slavery: The Master Class in Georgia and South Carolina, 1670–1837* (Chapel Hill: University of North Carolina Press, 1999).

15 See Mart A. Stewart, "Let Us Begin with the Weather?". U.B. Phillips, in *Life and Labor in the Old South* revived this argument in a twentieth century version that also converged with arguments about geographical determinism that were afoot at the time he was writing. See above, fn. 2.

Those who worked the land and the understandings they developed and employed as agricultural workers were as important to the environmental history of the plantations as was the structure of agriculture and of crop regimens—no consideration of the agroecology of Southern plantations can be extracted from a discussion of the labor system that drove it. Much of the South was shaped by the production of a very few staple crops on plantations, but more directly by the laborers who grew these crops. As Philip Morgan and Ira Berlin have pointed out, cultivation and culture were always linked in the plantation South and Caribbean; how people *worked* tells us a great deal about their cultures. Morgan and Berlin emphasize labor much more than land and the work culture of slaves more than the complex set of relationships they had with the environment. The hands that shaped plantation agriculture also shaped their own countervailing and sometimes competing landscapes, however, because of and by way of the work that they did on the land.[16]

Slave Landscapes and African American Conservation

In plantation districts, both the cultivated and uncultivated environments were often better known by slaves than by their masters. The work slaves did accustomed them to a closer view of the cultivated environment. They were aware, from row to row, of the progress of the plants during the growing season. They put seeds in the ground and covered them with their feet, stirred and tilled the earth when hoeing, and bent down over rice stalks or moved slowly down rows of cotton during harvest. The hands experienced crop cultures from the ground up. Masters sometimes even depended on the first-hand—and often more tangible—perceptions of leading slaves to make decisions about crop regimens. At the same time, when a storm came up slaves

16 For an analysis of how labor was the nexus of culture and cultivation in slave societies, see the essays in Ira Berlin and Philip D. Morgan, eds., *Cultivation and Culture: Labor and the Shaping of Slave Life in the Americas* (Charlottesville: University Press of Virginia, 1993). That not just "work" and labor needs to be examined to understand plantation agriculture and culture, but also *labor on the land* is one of the core arguments of Stewart, *"What Nature Suffers to Groe"*. See also Stewart, "Rice, Water, and Power: Domination and Resistance in the Low Country, 1790–1900," *Environmental History Review* 15 (Fall, 1991): 47–64.

went in the fields or out on the levees or rice banks to do repairs and salvage crops. They endured suffocating heat—especially in the low country rice swamps or in the damp thickets of Lower Mississippi sugar plantations—while doing the heavy labor of tending and harvesting the crops. Masters and overseers rode or strolled along the borders of the fields and sometimes down the rows, but the slaves who turned the soil, tended the plants, and harvested the crops acquired a first-hand knowledge of the cultivated landscape on the plantation.[17]

Slaves knew the woods and swamps that were not cultivated, too, and often as intimately. The conduits and seams of significance in slave landscapes were marked out not by the boundaries of the fields they were forced to work, but by the pathways and waterways along which they acquired opportunities for small measures of autonomy beyond the fields. They met in the holler for worship, and many depended in part on the local environments for sustenance, oak or seagrass for baskets, roots and herbs for medicine or other purposes—even quilt patterns.[18] Hunting and fishing in the surrounding woods and waterways were an important source of food for slaves. Not all slaves hunted—some plantation surroundings were not rich enough in game to yield much to hunters, and going off the plantation without a pass was too risky in some neighborhoods. But many did, if not with the rare guns they were able to use as hunters for their masters or that they owned themselves, with an ingenious array of snares, set traps and turkey pens. Or whatever else was at their disposal: Georgian Aunt Harriet Miller reported to a WPA interviewer that when she was a slave, she and other slaves used blow guns made out of sugar cane and burned out at the joints to

17 On slave perceptions of agricultural work, see Stewart, *'What Nature Suffers to Groe,'* 98–102, 135, 146–48.

18 Dale Rosengarten, *Row Upon Row: Sea Grass Baskets of the South Carolina Lowcountry* (Columbia: University of South Carolina Press, 1986); Leland Ferguson, *Uncommon Ground: Archaeology and Early African America, 1650–1800* (Washington: Smithsonian Institution Press, 1992), 59–120. The Works Progress Administration Slave Narratives, published in George P. Rawick, *The American Slave: A Composite Autobiography* (Westport, Conn.: Greenwood Publishing Co.), are rich—in some volumes, nearly every account yields something—about what slaves grew, hunted, gathered, or made from garden patches or edge places on the plantation and from all kinds of environments off. A useful introduction to some of these that also provides an analysis of the gendered quality of slave environmental practices, but that is far from exhaustive, is Elizabeth D. Blum, "Power, Danger, and Control: Slave Women's Perceptions of Wilderness in the Nineteenth Century," *Women's Studies* 31 (2002): 247–265. My thanks to Vongphone Luangphaxay for research assistance on this point.

"kill squirrels and catch fish."[19] With sometimes nothing more than motivation, opportunity, and a good stick, slaves sought something of their own by way of hunting. Slaves hunted everything, but the most common animals that found their way into pots in the quarters were opossums, raccoons, and rabbits. Rabbits were plentiful and had savory meat, roasted raccoon was meat with character, and the meat of the opossum, when scalded, rubbed in hot ashes, and roasted, and then eaten with roasted sweet potatoes and coffee, was prized most of all by slaves who hunted.[20] But whatever the animal, slaves had to be doubly stealthy and more knowledgeable than common for white hunters: they had to avoid stepping into their masters' landscapes of control and domination at the same time that they had to be closely attentive—especially if they were hunting merely with sticks and smarts and at night—to the nuances of the behavior and environment of their prey. Hunting put meat in the pot: On the Georgia and South Carolina coasts, for example, slaves may have procured nearly half the meat in their diets from wild sources—a crucial margin that added substantially to nutrition and sustenance.[21] At the same time, hunting was one more way that slaves acquired knowledge about the physical environment in their neighborhoods and annotated their surroundings with meanings that were both subversive of the totality of white power and positive expressions of an African American environmental ethos.

Again, what happened in the woods was linked to the interstices of agricultural regimens—and the history of plantation agriculture in the South. Most slaves devised ways to carve out some of their "own" time to expand their exploitation of local resources beyond the fields or apply specialized skills off task to cultivate, hunt, or gather after their work in the fields was done. Slaves were not only able to supplement rations and feed their families and neighbors. The food that slaves procured from the wild environment became imbued with cultural value when slaves developed a cuisine,

19 George P. Rawick, *The American Slave*, Georgia, vol 13, part 3, 130.

20 Nicolas Proctor, *Bathed in Blood: Hunting and Mastery in the Old South* (Charlottesville: University Press of Virginia, 2002), 144–168. Notes on the value of various animals for food can be found, for example, in Rawick, *The American Slave,* ser. 2, vol. 12, Georgia, pt. 1, 3–4; and supplement 1, Miss., v. 8, part. 3, 1293. See also Stuart A. Marks, *Southern Hunting in Black and White: Nature, History, and Ritual in a Carolina Community* (Princeton: Princeton University Press, 1991).

21 Tyson Gibbs, Kathleen Cargill, Leslie Sue Lieberman, and Elizabeth Reitz, "Nutrition in a Slave Plantation: An Anthropological Examination," *Medical Anthropology* 4 (Spring 1980): 175–262.

tastes for certain wild foods, and used gifts of meat and other foods to reinforce community bonds. They also used what they raised and procured in the wild places to trade for goods and property of their own. Cattle and hogs that ranged in the woods were, indeed, capital on the hoof, which increased by way of the browse that could be found there. Like their masters, slaves extracted commodities from the environment in which they lived and worked, and indeed masters often encouraged some property ownership by slaves—they believed it would make them less likely to run away, and sometimes slave property substantially supplemented plantation rations. Whatever property they could acquire had more than pure economic value, however. In a relationship with other humans and larger institutions that defined them as human property, outside civil society and subject to the almost absolute domination of their masters, small bits of property represented considerable increments of independence and autonomy, even when they also served the goals of masters. Property was not simply wealth, but represented a small measure of security and something that was the slaves' own, and more slaves than not had some.[22]

At the same time, natural resources and the property made from them were used not merely to strengthen individual positions of power, but were important in consolidating family bonds. Wild resources and the process of procuring them did not produce family, but were often the medium of kinship. Cooperative arrangements that freed some slaves to cultivate their own plots, fish, hunt, or gather and then trade or sell, were usually kin arrangements. Slaves worked with relatives to extract resources, relatives took care of property when the owner was absent, and some slaves got their start—a few chickens or a shoat or a calf—by way of a gift or a loan from a relative. When slaves disputed ownership of something, they negotiated a resolution by way of kinship networks—relatives or reliable neighbors were witnesses and trusted ones were arbiters. When slaves died, their children inherited what they had. The resources enslaved African Americans were able to gather

22 On the cultural meanings of slave food, see Charles W. Joyner, "Soul Food and the Sambo Stereotype: Foodlore from the Slave Narrative Collection," *Keystone Folklore Quarterly* 16 (1971): 171–178; Stacy Gibbons Moore, "'Established and Well Cultivated': Afro-American Foodways in Early Virginia," *Virginia Cavalcade* 39 (1989): 70–83; Philip D. Morgan, *Slave Counterpoint: Black Culture in the Eighteenth-Century Chesapeake and Lowcountry* (Chapel Hill: University of North Carolina Press, 1998), 134–45. On slave property, see Dylan C. Penningroth, *The Claims of Kinfolk: African American Property and Community in the Nineteenth-Century South* (Chapel Hill: University of North Carolina Press, 2003), 45–78.

or the small property they were able to procure because of these arrangements reinforced and further strengthened kinship ties. Property ownership was so interrelated with kinship for slaves that the making of property and the making of family often went hand in hand. Slaves metabolized resources from the fields, forests, and swamps of plantation neighborhoods in their social arrangements as well as adding to their food supply and nutrition. They crafted expressions of culture and values, and also quite literally claimed family ties with what they extracted and metabolized—both in the process and the product—from the environment.[23]

23 Ibid., 79–109. For slaves and later for freedmen, property was always connected to family, Penningroth argues, and "was less an institution or a legal right than a social process." He connects this insight to scholarship in African Studies that argues that access to resources is connected to social identity and that property ownership is more an ongoing social process than a matter of having something to the exclusion of the claims of others: Ibid., 191–92. The animal and trickster tales slaves told were comprehensive expressions of slave relationships to nature, negotiation, and kin: These tales were also a vehicle for the portrayal of slaves' perceptions of "natural" social relations in two senses. African Americans saw themselves as part of a unified universe of all creatures and did not make a sharp distinction between humans and other creatures. At the same time, these tales, especially the trickster tales, were depictions of social relations as the African Americans believed they were inscribed in nature. When a weak animal defeated a strong one by using its wits, this was a conquest with doubly meaningful social resonance. See Stewart, *"What Nature Suffers to Groe,"* 178–80. Such tales were common at least in South Carolina and Georgia. See Georgia Writers Project, *Drums and Shadows*, 79, 110–11, 160–61, 171. An older collection, with no notes on informants, is Charles C. Jones, Jr., *Negro Myths from the Georgia Coast Told in the Vernacular* (New York: Houghton, Mifflin and Co., 1888; repr. University of Georgia Press). Patricia Jones Jackson describes the discernment of distinct features of particular animals that sea island storytellers bring to their tales: *When Roots Die: Endangered Traditions on the Sea Islands* (Athens: University of Georgia Press, 1987), 16–17, 171n–72n. African Americans also developed a strong sense of place, which wove together networks of kin and their close understanding of local environments. Though many moved around after Emancipation to reunite with kin and to escape the supervision of their ex-masters, Freedmen's Bureau officials who worked with freedmen and women after the Civil War often noted the strong loyalty to place and the resonance of place with kinship networks that many emancipated slaves continued to express: Herbert G. Gutman, *The Black Family in Slavery and Freedom, 1750–1925* (New York: Pantheon Books, 1976), 208–11; Drew Gilpin Faust, "Culture, Conflict, and Community: The Meaning of Power on an AnteBellum Plantation," *Journal of Social History* 14 (Fall 1980): 93–94; Patricia Guthrie, "Catching Sense: the Meaning of Plantation Membership Among Blacks on St. Helena Island, South Carolina" (Ph.D. dissertation, University of Rochester, 1977), 114–29; Steven Hahn, *A Nation Under Our Feet: Black Political Struggles in the Rural South from Slavery to the Great Migration* (Cambridge: Harvard University Press, 2003), 139–40.

Black Southerners who knew how to extract resources discretely and who occupied a natural landscape that was leavened with strategies for strengthening kin and community had the makings of a different environmental ethos that also operated in tension with the conservation ethós of their masters—and that also shaped plantations as agroecological enterprises in general. But it was this very experience with the conservation ethic and other demands of elite Southerners—those who owned them but to whom they were partly invisible—that contributed the crucial element to African American environmentalism (even as it has bloomed in more recent times). Slaves were required to negotiate for everything, either directly and indirectly, with masters and with the systems of control they devised. They had to bargain with both words and behavior for access to resources, to move around on the plantation and beyond the bounds of the cultivated fields, to manipulate adjustments to the burden of labor that was placed upon them, and to do all in the interest of kin and community. Anything they did for themselves was potentially and sometime quite overtly an act of resistance, and had to be negotiated carefully. Even the medicine they sought to apply to treat illness, even if it brought back a slave's health and his or her capacity to be a productive worker, was usually regarded by planters, who sought to control the bodies of slaves as well as what those bodies could do, as an act of subversion. Reformers and Freedmen's Bureau officials who worked with freedmen in the South just after Emancipation were often surprised—stunned, even—by the speed and deftness and collective force with which freedmen laborers negotiated with landowners or managers to mark out better terms for themselves. They remarked often about the rapidity with which freedmen and women organized churches—usually with denominational lines that follow kinship and neighborhood ones—that also became homes to community political activities and expressions. What they were witnessing and experiencing was not something new, but a political behavior with deep roots in the conditions of American slavery and in the relationship of African Americans to the land.[24]

24 See Hahn, *A Nation Under Our Feet*, 2, 128; Jack Temple Kirby, *The Countercultural South* (Athens: University of Georgia Press, 1995), 8–32. Kirby argues that deep traditions of negotiation and collective action have given African Americans more political power, once segregation was demolished, than poor whites—who have tended to withdraw into isolation or into individual acts of subversion. This argument has not gone unchallenged, by the way—mainly because it tends toward an interpretation that sees the "slave community" as a unified organic unit—a notion historians no longer accept.

Plantation Legacies

This history left postbellum black Southerners with a double-edged inheritance. Those who lived in the old plantation districts were more likely, in the twentieth century, to live in poverty than their urban African American counterparts. Again, poor, underdeveloped counties in the South with large black populations have also been more likely to be locations or proposed locations of hazardous waste sites or factories that spew noxious pollutants. But slavery and emancipation and the political culture that came out of them—both in the countryside and in the urban places to which rural southern blacks migrated—have produced a positive response to injustices—environmental ones included. Relationships with the environment have always been social and collective for African Americans, and always in process of negotiation—and the site of these negotiations was also the place where African Americans were subjected to perpetual bondage, the plantation.[25]

Nature on and around the plantation provided resources not just for profit but often to consolidate community—moving into nature and through nature was usually a collective matter, as was negotiating either individual or group spaces and access to resources from masters. For African Americans, both cultivated and uncultivated places on and around plantations were sites of healing, a trail to kinship, a place where a decisive edge of resources could be added to meager plantation rations, a place where salvation could be gained—either through worship in the holler or through stealing oneself away. Slave experiences with the environment were profoundly social and political ones—they moved into nature to enact social meanings, at the same time that they did not make the sharp distinction between the human and nonhuman worlds that were common for whites. For African Americans in the South, nature was negotiated, it was kin, and it was community.

25 Charles S. Aiken, *The Cotton Plantation South Since the Civil War* (Baltimore: Johns Hopkins University Press, 1998), 360–61. Steven Hahn explains how rural black Southern culture was transported to urban places during the Great Migration, and there became the foundation for Garveyism and other important expressions of collective action and black nationalism: Hahn, *A Nation Under Our Feet*, 465–78. For a review of the origins of African American environmentalism, see Mart Stewart, "Nature, Negotiation, and Community: Slavery and the Origins of African American Environmentalism," in *"To Love the Wind and the Rain": Essays in African American Environmental History*, eds. Dianne Glave and Mark Stoll (Pittsburgh: University of Pittsburgh Press, 2005); and Mart A. Stewart, "If John Muir Had Been an Agrarian: American Environmental History West and South," *Environment and History*, 11(2) (May 2005): 139–162.

Plantations in the U.S. South, then, were sites of both intensive and extensive agriculture, places both of intense environmental manipulation and of uncultivated edges, complex assemblages of monoculture fields and polyculture patches. They were also sometimes sites of agricultural experimentation, productive nodes in the global circulation of environmental knowledge, and places where social and political relations were naturalized. The environmental or conservation ethics that planters and slaves developed in tandem can be regarded as antecedents for 20th century soil conservation initiatives and the environmental justice movement, respectively. But seeing them mainly in these terms fails to historicize them properly—they do not really make sense in these terms, in fact. And though the story of soil conservation in the antebellum South is a story of failure and the story of African American environmental ethics is a story of mere survival, looking at them on their own terms contributes to our understanding of the varied global history of plantations.

Cotton's Metropolis: Memphis and Plantation Development in the Trans-Mississippi West, 1840–1920

Jeannie M. Whayne

As the cotton plantation moved west in the antebellum period, Memphis, Tennessee, became the largest inland cotton market in the region. Even as cotton cultivation shaped the destiny of Memphis, the city's businessmen and cotton factors encouraged the further exploitation of the landscape in western Tennessee, northern Mississippi, and eastern Arkansas, prompting deforestation and, eventually, massive drainage enterprises that destroyed wetland habitats. The demand for cotton abroad played a crucial role in stimulating this exchange between city and countryside, between Memphis and the world cotton market. Its location along the Mississippi River guaranteed that Memphis would become a nexus of the international trade in cotton, the commodity that was at the heart of the British textile industry, and a focal point of global capitalism. The South enjoyed a monopoly of the cotton trade in the antebellum period, and Memphis, as the emerging center of the fastest growing cotton region, benefited from this dominance. But the city and the cotton economy both would face serious problems in the 19th and early 20th centuries. The Civil War, by internationalizing the cultivation of cotton presented challenges, and although southern cotton producers would regain their dominance in the world cotton market, prices would only very briefly return to pre-war levels. Too many competitors fundamentally altered the profit structure. To add insult to injury, yellow fever devastated the city and malaria undermined the productive capacity of individuals in the entire Memphis region. The cotton producers in the city's orbit also faced environmental problems that necessitated a new relationship with government. Increasingly severe floods along the entire lower Mississippi River valley led to calls for flood control measures, creating the possibility for the coalescence of interests developing around the issue, linking the federal government, science, and the cotton elite in the Memphis area.[1]

1 James David Miller, *South by Southwest: Planter Immigration and Identity in the Slave South* (Charlottesville: University of Virginia Press, 2002). Mart A. Stewart, *"What Nature Suf-*

According to Memphis Cotton Exchange promotional materials, the Memphis trade area included northern Mississippi down to Greenville, a large segment of eastern Arkansas, western Tennessee, and southeast Missouri, particularly the boot heel.[2] For the purposes of this analysis, the area identified as within the Memphis orbit includes seven counties in eastern Arkansas, fifteen counties in west Tennessee, and sixteen counties in north Mississippi.[3] A perusal of the figures for improved acres in production and the number of cotton bales produced in the area establish four crucial periods (see Tables I through IV). First, despite a poor reputation for disease, Memphis grew from an insignificant village in the mid-1830s to a major cotton entrepot in the late antebellum period, driven mostly by the expansion

fers to Groe": Life, Labor, and Landscape on the Georgia Coast, 1680–1920 (Athens: University of Georgia Press, 2002); Mikko Saikku, *This Delta, This Land: An Environmental History of the Yazoo-Mississippi Floodplain* (Athens: University of Georgia Press, 2005); Lynn A. Nelson, *Pharsalia: An Environmental Biography of a Southern Plantation, 1780–1880* (Athens: University of Georgia Press, 2007); Cynthia Barnett, *Mirage: Florida and the Vanishing Waters of the Eastern U.S.* (Ann Arbor: University of Michigan Press, 2007); Michael Grunwald, *The Swamp: The Everglades, Florida, and the Politics of Paradise* (New York: Simon & Schuster, 2006); David McCally, *The Everglades: An Environmental History* (Gainesville: University Press of Florida, 1999). For an overview of cotton's importance in the American—not just southern—economy from the antebellum period forward, see Gene Dattel, *Cotton and Race in the Making of America: The Human Costs of Economic Power* (Chicago: Ian R. Dee, 2009). See also an interesting treatment of the worldwide importance of cotton production in the mid-nineteenth century by Sven Beckert, "Emancipation and Empire: Reconstructing the Worldwide Web of Cotton Production in the Age of the American Civil War," *American Historical Review* 109, no. 5 (December 2004): 1405–1538. Beckert's manuscript on the role of cotton in the American and world economy is due out in 2011. For a different take on the convergence of interests that emerged to control water—this time to harness it for irrigation in the west—see Donald Worster, *Rivers of Empire: Water, Aridity, and the Growth of the American West* (New York: Oxford University Press, 1985).

2 Lynette Boney Wrenn, *Crisis and Commission Government in Memphis: Elite Rule in a Gilded Age City* (Knoxville: University of Tennessee Press, 1998), 4. Wrenn characterizes this area as the city's "commercial hinterland."

3 The Arkansas counties are Crittenden, Cross, Lee, Mississippi, Phillips, Poinsett, and St. Francis; the Mississippi counties are Benton, Bolivar, Carroll, Coahoma, DeSoto, Grenada, Lafayette, Leflore, Marshall, Panola, Quitman, Sunflower, Tallahatchee, Tate, Tunica, and Yalobusha; the Tennessee counties are Carroll, Chester, Crockett, Dyer, Fayette, Gibson, Hardeman, Haywood, Henry, Lauderdale, Madison, Obion, Shelby, Tipton, Weakley. While the Tennessee River flows through northern Mississippi, by the time that the Chickasaw lands were purchased by the United States in 1834, Memphis was already established and though the Tennessee River was important, no town of significance developed capable of challenging Memphis' hold over the cotton trade in the region.

of the cotton plantation in northern Mississippi, especially in the 1850s. The second period (1860 to 1880) encompasses the Civil War interlude, which temporarily interrupted the expansion of the plantation—just as it disrupted production of cotton on the plantations in the older South. Yet Memphis, a garrisoned city, held its own, and emerged from the war relatively unscathed. However, the war also made possible the development of fierce competitors in the form of cotton producers in Brazil, Egypt, and, particularly, India, and made recovery of the cotton economy in the American South distinctly problematic. By 1880, despite serious financial constraints, the American cotton producers were on their way to regaining their share of the global cotton market, but prices would only briefly match their pre-war levels. Significantly, the Memphis cotton factors and businessmen were able to use the chaos arising out of a yellow fever epidemic in 1878 to reclaim political and economic ascendancy in the city, something they had lost to a coalition of black and Irish voters. By the third period (1880 to 1900), the southern cotton producers had reclaimed their dominant position in the now worldwide production of cotton. Cotton production in the Memphis region set the pace, particularly with expanded acreage in Tennessee and Mississippi. But cotton prices remained erratic and floods along the Mississippi River endangered the Memphis harbor and began to challenge those who settled on its banks. It was in this period that the marriage between government, science (engineers) and the cotton elite was forged. The fourth period (1900 to 1920) included a surge in production in the Arkansas delta, such that by 1920 it rivaled west Tennessee in terms of cotton bales produced. Successful drainage enterprises enabled Arkansas planters, for example, to reclaim hundreds of thousands of additional acres in this period, but much of the swamplands remained, and a new and deadly variety of malaria made its appearance, threatening the lives of the region's residents. According to an early 20th century government report, the malaria rate in the Memphis area was higher than any other city in the lower Mississippi River valley.[4] Clearly, there was no point between 1840 and 1920 when Memphis was without challenge.

4 American Association for the Advance of Science, Bulletin 15, "A Symposium on Human Malaria with Special Reference to North America and the Caribbean Region" (Smithsonian Institution, Washington, D.C., 1941), 10.

Antebellum Era, ~1834 to 1860

An observer in 1830 might have doubted, quite reasonably, whether the small hamlet of Memphis could develop into the region's cotton entrepot. It was burdened with a well-deserved reputation as a "sickly" place and vied with another town just north of Memphis, Randolph, as a destination for the many barges and packets traversing the river. The expansion of the cotton economy in western Tennessee in the late 1820s and early 1830s made necessary a marketing center on the river, and enterprising individuals settled in the two locations and began to lay plans to promote their respective sites. Both towns were situated on one of the four Chickasaw bluffs—Memphis on the fourth, Randolph on the second—which provided some protection against the Mississippi River. A geographic feature which stretches from Lauderdale, Tennessee to Memphis, the bluffs stand fifty to one hundred feet above the Mississippi River flood plain. During high water periods, the river pressed eastward into the bluffs but then, taking the least line of resistance, flooded westward into the lowlands of Arkansas. The Chickasaw Bluffs were not impervious to the river's force, however. Composed of loess (which is a blown sand and silt composition) over a gravel base, they can erode, particularly given the constant pressure of the fast moving and powerful Mississippi River.[5]

The two towns competed fiercely to become the region's entrepot, but several factors converged to turn the advantage to Memphis. Founded in 1819 in Shelby County, just a year after the purchase of the Chickasaw lands in western Tennessee, Memphis was heavily promoted by Andrew Jackson and two other prominent Tennesseans. Designated the Tipton county seat, Memphis was assured the important legal business in the area. Randolph, founded in 1823, vied to become the county seat there but lost that designation to Covington, an interior town along the Hatchie River. Another blow to Randolph boosters occurred when Memphis secured the tri-weekly mail service from the state's capital, Nashville. Jacksonian era Indian removal policies further benefitted the town. The Treaty of Pontotoc in 1832 appropriated Chickasaw lands in northern Mississippi, thus guaranteeing the removal

5 Floyd M. Clay, *A Century on the Mississippi: A history of the Memphis District, U.S. Army Corps of Engineers, 1876–1981* (Memphis: U.S. Army Corps of Engineers, 1886), 33–35.

of the Chickasaw and providing Memphis with access to some of the richest land in the region.[6]

Memphis also held a geographical advantage for those hoping to promote it as a commercial center. Although the Hatchie River, which ran 238 miles from Mississippi northward before turning west and passing into the Mississippi at Randolph, had the potential for moving commerce from those rich northern Mississippi and western Tennessee cotton-producing interior counties, Randolph was further from the source of much of the cotton. Memphis was situated where the Wolf River entered the Mississippi, and although it was half the length of the Hatchie, it too ran from Mississippi, providing a conduit from the rich cotton counties soon to develop there, and from the perspective of the northern Mississippi cotton producers, it was in easier reach. Perhaps even more significantly, the Mississippi River itself chose Memphis, although the river's favor was a mixed blessing, guaranteeing the town would face continuing—and increasingly severe—environmental difficulties. The Mississippi River runs from the northwest in a southeasterly direction in an almost straight line toward Memphis, slams into the bluff and then turns sharply to the southwest before straightening out, more or less, and moving south to New Orleans. Randolph, situated just to the north of Memphis, would see the river slowly move away from its environs toward the Arkansas bottomlands. But the force of the river crashing into Memphis also undermined the Chickasaw bluffs and created other difficulties that would eventually necessitate extraordinary measures, the very measures that only a combination of government bureaucrats and engineers—at the urging of the cotton elite—could accomplish.

The environmental consequences of the river's trajectory were not immediately recognized, and as cotton cultivation and the accompanying development of the slave plantation system in the Memphis orbit vastly expanded in the 1840s and 1850s, Memphis became the great riverboat city in the region. Spurred by a significant increase in cotton prices in the 1850s, planters cleared and cultivated an additional 611,541 acres in northern Missis-

6 "History and Facts about Memphis and Shelby County", Memphis Public Library, accessed March 18, 2012; Marshall Wingfield, "Town of Randolph: Turbulence Marked Its Brief Career," *Memphis Commercial Appeal*, November 27, 1949 (Memphis Information Files, Randolph, Memphis/Shelby County Public Library and Information Center, Memphis, Tennessee). See also Don Wilson, "Randolph, The Glory Years," *The West Tennessee Historical Society Papers*, Vol. LI (December 1997): 98–105; and Gerald M. Capers, Jr., *The Biography of a River Town, Memphis: Its Heroic Age* (Chapel Hill: University of North Carolina Press, 1939), 57–59.

sippi, western Tennessee, and eastern Arkansas during that decade, and the number of slaves increased from 146,035 to 203,111, a 39 percent increase (see Tables I and II).[7] The slave population and cotton cultivation increased most significantly in northern Mississippi on the old Chickasaw lands. Based on the number of bales marketed (the census did not report the acreage devoted to cotton until 1880), it is clear that planters and farmers devoted much of the new acreage in Mississippi and Tennessee to cotton. Through the use of substantial numbers of slaves, the planters in the region cultivated 99,630,000 pounds of cotton in 1849 (see Tables III and IV). A decade later they had more than doubled that amount, harvesting a total 233,139,500 pounds in 1859. Most of this increase occurred in the Mississippi counties which now grew 59.3 percent of the cotton harvested in the three areas. Tennessee, which harvested more cotton than it had in 1850, had declined in importance as a grower (from 45.2 percent to 33.3 percent) compared to Mississippi and Arkansas. The latter had quadrupled its cotton production but remained far behind in terms of the absolute number of bales of cotton marketed though it had increased its share (from 3.3 percent to 7.4 percent). The swamps of northeastern Arkansas discouraged intensive development in that region for some decades more, though even there large plantations associated with a prominent Nashville planter family, lined the river in Mississippi County, Arkansas. Protected by a natural levee but still subject to overflows, the planters of northeast Arkansas remained compromised in their ability to fully develop the land west of the Mississippi River.[8]

By 1860 Memphis had gained its place as the largest inland cotton center in the United States, and given the fact that the demand for cotton by the textile mills in the north and in England appeared to be limitless, prospects for further growth seemed assured. "On the eve of the American Civil War in the mid-1800s cotton was America's leading export, and raw cotton was essential for the economy of Europe."[9] As the sectional crisis intensified in the late 1850s, secessionist forces in all three states grew stronger. None of them could imagine how they could maintain their production levels without

7 The number of slaves, in fact, increased at a greater rate (39 percent) than the number of acres in production (33.5 percent), signaling, perhaps, that many of those new slaves were involved in reclaiming land, an arduous and time-consuming endeavor.

8 Lonnie Strange, "Civil War and Reconstruction in Mississippi County, Arkansas: The Story of Sans Souci Plantation" (Honors Thesis, Spring 2008, University of Arkansas, Fayetteville), 8.

9 Gene Dattel, "Cotton in the Civil War," http://mshistory.k12.ms.us/articles/291/cotton-and-the-civil-war (accessed June 17, 2011).

slavery, and Lincoln's anti-slavery rhetoric seemed to portend disaster. The economic and political realities in the states of Arkansas, Mississippi, and Tennessee, however, took them on different paths toward secession in 1861. After South Carolina seceded on December 1860, Mississippi, which had the highest percentage of slaves (54 percent) of the three states in the Memphis orbit and a legislature dominated by the planter faction, quickly followed in January. Arkansas and Tennessee, with 25 percent of their population enslaved, delayed until after the firing on Fort Sumter. Secession proved to be no protection for their prosperity, however. The South had badly miscalculated on at least two fronts: the influence of their economic position with respect to cotton; and the determination of the new president to maintain the union.[10]

1860–1880

The period between 1860 and 1880 presented the city with unprecedented challenges. Not only did it endure military defeat and occupation during the Civil War, Memphis also faced a disastrous epidemic in 1878 that dealt an almost fatal blow. In addition to undergoing an economic catastrophe from the collapse and transformation of the cotton economy, the city underwent

10 Arkansas, with a strong non-slaveholding sector existing in the northwestern part of the state, declined to secede in late January 1861, choosing instead to put the matter to a vote of the people scheduled for August. After South Carolinians fired on Fort Sumter in April, however, Arkansans called its secession convention back into session in May and voted almost unanimously to secede. In Tennessee, which like Arkansas had 25 percent of its population enslaved, the people voted against a secession convention, but when Lincoln called for troops to put down the rebellion after Fort Sumter, the state's governor called for a secession referendum. On June 8, 1861, despite a strong anti-secession contention in east Tennessee, voters in the middle and western parts of the state, the areas most devoted to cotton and slaves, carried Tennessee out of the union, making it the last state to secede. In all three states, the most outspoken secessionists, who lived in the areas of the state most dominated by slavery and cotton, rallied the state to the secession cause. For Arkansas and secession, see Carl Moneyhon, *The Impact of the Civil War and Reconstruction on Arkansas: Persistence in the Midst of Ruin* (Baton Rouge: Louisiana State University Press, 1994); for Tennessee, see Stephen V. Ash, *When the Yankees Came: Conflict and Chaos in the Occupied South, 1861–1865* (Chapel Hill and London: The University of North Carolina Press, 1995); and for Mississippi, see Clay Williams, *The Road to War, 1846 to 1860,* http://mshistory.k12.ms.us/articles/206/the-road-to-war-1846–1860 (accessed June 20, 2011).

a political revolution that greatly reduced the influence of the business and cotton elite, and put into power a petit bourgeoisie made up of small businessmen, many of them Irish, who became masters at ward politics. At no other time in its history has Memphis undergone such a dramatic transformation, but it was also at the end of this era that the cotton elite regained control and began to put into place economic and structural reforms that made Memphis over again.

Given the importance of cotton to the economy of the United States and the textile industry in England, it is no wonder that the planters in the Memphis orbit, like those elsewhere in the South, had an inflated sense of their own importance in 1860. Cotton accounted for nearly 60 percent of U.S. exports by 1860 and enriched the New York financiers who formed an important link in the chain of trade with Britain. Cotton planters believed that these New Yorkers would be forced to side with them after they formed the Confederacy and began practicing "cotton diplomacy." Although many of those eastern financiers were unenthusiastic about the conflict and its implications for their businesses, few of them offered much support beyond vague phrases of sympathy. Although they formed an important source of support for Democratic politicians who criticized Lincoln's war policy, this did not contrive to work much benefit to the southerners. Neither could the South find it possible to convince England to provide diplomatic recognition. Bumper crops in the late 1850s had allowed England to stockpile cotton and thus it could bide its time, sensing no urgency in giving the South the recognition it so desperately wanted. English textile mill owners successfully influenced the development of new colonial policies that vastly expanded cotton production, particularly in India, a fact that would dampen England's enthusiasm for intervention in the American conflict, transform the world cotton market, and have long term implications for southern cotton producers.[11]

While the Lincoln administration imposed an increasingly effective blockade in order to keep the Confederacy from exporting its cotton, traders managed to slip by the federal Navy and reach Nassau and from there the cotton was transported to its destination in England. Memphis would become a center of the illicit trade, but much of the cotton grown in its orbit was made to appear legitimate—that is, not grown by nor accruing to the

11 Beckert, "Emancipation and Empire"; Dattel, *Cotton and Race in the Making of America*. See also Graeme J. Milne, *Trade and Traders in Mid-Victorian Liverpool: Mercantile Business and the Making of a World Port* (Liverpool: Liverpool University Press, 2000).

interest of the Confederacy. Steamboats from Memphis transported cotton up river, delivering it to the cotton-starved textile mills in the north and on to Great Britain. Given that Memphis was occupied by the Union military for most of the war, the cotton trade was subjected to the eager attention of the occupiers, and by late 1862 Memphis cotton purchasers had to have "special permits from both the Treasury Department and the Provost Marshal, and all the purchases [had to be] made within the city." This meant the Confederate cotton had to pass through union picket lines, but some union sentries were themselves complicit in the illegal trade in Confederate cotton. According to one government report, "Every colonel, captain, and quartermaster is in secret partnership with some operator in cotton; while every soldier dreams of adding a bale of cotton to his pay." This is almost certainly an exaggeration, but with cotton reaching nearly a dollar a pound in early 1863, the temptation was likely too great for some. At a dollar a pound, a bale of cotton would have netted $500, just a hundred dollars less than the average annual income in the United States in 1860.[12]

While the Civil War interrupted the growth of cotton cultivation in the Memphis trade area, the city was saved from the disasters faced by other southern cities. Because of Union occupation for most of the war, its buildings and homes were spared destruction. Captured by federal forces after the Battle of Memphis on June 6, 1862, which gave them control of the river down to Vicksburg, Memphis became an important staging ground for union military activities down river. It remained heavily garrisoned throughout the war, and although merchants in Memphis were damaged by shifting policies that periodically restricted and liberalized trade with the interior in a futile attempt to keep cotton profits out of the hands of Confederates, the town's merchants were not as injured as others elsewhere. The city's businessmen were sitting in the heart of one of the largest cotton producing areas in the region, and by 1863 a policy of leasing plantations to either "loyal" southerners or to northerners had been adopted in western Tennessee and in small parts of Arkansas and northern Mississippi. However, historian Stephen V. Ash characterizes the region around garrisoned cities like Memphis as a "no man's land," where neither the protection of the Union army nor the Confederate army existed. Beyond the no-man's land was the Confederate frontier, another place where little protection could be guaranteed to citi-

12 Joseph H. Parks, "A Confederate Trade Center Under Federal Occupation: Memphis, 1862–1865," *Journal of Southern History* 7 (August 1941): 301. See also Capers, *The Biography of a River Town*, 153–54.

zens, including farmers and planters trying to plant and harvest crops. Cotton production in Mississippi dropped by half, from 138,950,000 in 1859 pounds to 65,144,000 in 1869 (see Table III). While Arkansas producers held on to their small share of the production in the region, cotton production in Tennessee was reduced by nearly 20 million pounds. Since these figures encompass the four years after the war ended, it is highly likely that the decline during the war itself was much more severe. Still, many planters and farmers managed to produce a crop on the edge and within Confederate territory, and much of the cotton that came through Memphis was Confederate-grown and potentially benefited the Confederate war effort. Federal authorities struggled without success to regulate the cotton trade in a frustrating attempt to keep profits out of the hands of Confederates.[13] By 1863, according to one estimate, "$12,000,000 worth of imports passed through Memphis to the Confederate armies within a period of eight months" and much of this would have been purchased by virtue of cotton production.[14] Once federal officials realized that even loyal citizens and sometimes their own officers and men intended to realize the profits to be had in trading cotton and other items, they authorized the commanding officer in the city to adopt policies that were detrimental to the Memphis cotton traders. By 1864, the town was effectively closed to trade.[15]

Memphis survived the Civil War intact, but, like the rest of the cotton South, suffered by the failure of the cotton economy to fully revive. Improved acreage in production in the Memphis orbit barely changed between 1859 and 1869, rising from 2,434,810 to 2,462,931.[16] Cotton production decreased significantly, from 231,139,500 to 140,592,000 pounds. The stagnation in cotton acreage was typical across the South, as a matter of fact. Cotton production south wide decreased significantly in 1862 and would not return to the high it achieved in 1861 until 1870. As Sven Beckert suggests, however, the South was well on its way to recapturing its dominant position

13 Stephen V. Ash, *When the Yankees Came: Conflict and Chaos in the Occupied South, 1861–1865* (Chapel Hill and London: The University of North Carolina press, 1995), 77–78.

14 Historian Joseph H. Parks cites the New York *Herald*, quoted in Memphis, *Daily Appeal*, March 31, 1863, in Parks, "A Confederate Trade Center," 304.

15 Parks, "A Confederate Trade Center," 298–299.

16 The short crop in 1869 was due in part to weather conditions in the area during 1869, the year for which county level data is available. A sustained drought in the summer and too much rain at harvest damaged the crop.

as the world's cotton supplier, a remarkable feat for a region devastated by war and facing new competitors from abroad.[17]

	NATIONWIDE	MEMPHIS MARKET AREA
Year	*Bales*	*Bales*
1859	4,508,000	466,279 (10.3%)
1860	3,841,000	
1861	4,491,000	
1862	1,597,000	
1863	449,000	
1864	229,000	
1865	2,094,000	
1866	2,097,000	
1867	2,520,000	
1868	2,366,000	
1869	3,011,000	281,184 (9.3%)
1870	4,352,000	

The figures for cotton production in the left hand column derive from statistical abstracts and are not provided for discrete areas—like the Memphis market, but the census of agriculture for 1859 and 1869 (in the right hand column) demonstrate that Memphis had nearly held on to its percentage of the American market share by the end of the ten-year period. As difficult as conditions were for the Memphis cotton factors and merchants, the city "came through the war unscathed physically and its business institutions were reasonably intact. Atlanta, Richmond, and several other cities were destroyed. They literally had to start over from scratch in mid-1865."[18]

17 Beckert, "Emancipation and Empire." See also Dattel, *Cotton and Race in the Making of America*, who argues "By 1870, sharecroppers, small farmers, and plantation owners in the American south had produced more cotton than they had in 1860, and by 1880, they exported more cotton than they had in 1860." But the better comparison might be to look at 1861 when the South actually produced/exported 4,491,000 bales. Ibid.; Dattel, "Cotton in the Civil War."

18 Robert A. Sigafoos, *Cotton Row to Beale Street: A Business History of Memphis* (Memphis: Memphis State University Press, 1979), 44–45. Bureau of the Census, *Ninth Census, Vol. III, The Wealth and Industry of the U.S., 1870* (Washington: Government Printing Office,

Complicating the efforts of the city's business elite to effectively address the problem was the fact that they lost political control in the early 1870s to a coalition of Irish and African American citizens. Such an alliance would have seemed unimaginable just a few short years earlier when a mob, made up of native white southerners and Irish Americans, rampaged against the city's black population in an infamous race riot in 1866. But both groups of voters proved to be politically astute and after freedmen received the vote later in that same year, their leaders recognized that they could capture control of city government. While middle class Irishmen like Jack Walsh, an undertaker by trade, ruled the so-called "bloody first" ward, African American voters in the fifth ward were guided by Captain Billy Rice Brown. Otherwise, ward bosses came and went, and there was no "single city-wide political boss." The ward bosses heavily influenced the selection of aldermen, and the alderman "controlled municipal patronage and contracts." The business and cotton elite were forced to negotiate with them, but the cotton economy was not without importance to the middle-class businessmen in the city and the ward bosses understood the connection.[19] They were also in tune with the interests of their working class constituency, and many of them were tied to the cotton economy, either directly or indirectly. They loaded cotton at the wharf or the railroad yards; they worked in one of the large compresses; or they labored for commission merchants who supplied the merchants doing business in the countryside. "Of the forty-nine largest American cities in 1870, Memphis had the highest percentage of its workforce engaged in occupations related to trade and transportation."[20] The Irish, who constituted the largest foreign-born population in Memphis, made up 1,422 of the 3,284 working people in the city, competing with the impoverished black population for the jobs at the lowest level.[21]

1870), 100, 102, 184, 186, 244, 246, 245, 247; Bureau of the Census, *Ninth Census, Statistics of the Population of the United States, Vol. I* (Washington: Government Printing Office, 1872), 13–14, 41–43, 61–63.

19 Ibid., 25. For historiography on the race riot, see Altina Waller, "Community, Class, and Race in the Memphis Riot of 1866," *Journal of Social History*, vol. 18, no. 2 (December 1984): 233–246; Bobby Lovett, "Memphis Riots: White reaction to Blacks in Memphis, May 1865–July 1866," *Tennessee Historical Quarterly* 38 (1979): 9–33; and James G. Ryan, "The Memphis Riots of 1866: Terror in a Black Community during Reconstruction," *The Journal of Negro History*, vol. 62, no. 2 (1979): 243–257.

20 Wrenn, *Crisis and Commission Government in Memphis*, 2.

21 Gerald M. Capers, Jr., *The Biography of a River Town*, 107–108; Dernoral Davis, "Hope versus Reality: The Emancipation Era Labor Struggles of Memphis Area Freedmen, 1863–1870," in *Race, Class, and Community in Southern Labor History*, Gar M. Fink and Merle

The city's wealthiest citizens lived a world apart, not only in location but in the lavish lifestyle many of them enjoyed in spite of the economic woes confronting Memphis in the 1870s. They built elegant mansions, some of them occupied by planter families within the Memphis orbit, in the fashionable avenues south of downtown Memphis, particularly along Vance Street. They attended balls, frequented the Memphis symphony, formed exclusive clubs and associations, and otherwise withdrew into a tightly knit circle that included prominent businessmen, lawyers and judges, and prosperous cotton factors. Many of them, particularly the cotton factors, invested in plantation lands or, at the very least, had extensive business relationships with planters.

The relationship between the cotton factors and the planters—whether the latter chose to build mansions in Memphis or not—was an intimate one. Both before and after the Civil War, cotton factors in Memphis operated as commission merchants. They contracted with planters to cultivate cotton, took delivery of the crop upon harvest (or had an arrangement with a compress and storage firm to take delivery), and advanced funds to planters to put a crop in the ground and to furnish their necessities between harvests. In some ways, cotton factors resembled banks in how they managed the funds of planter families. For example, J. & J. Steele & Company, prosperous cotton factors doing business at No. 1 Exchange Building on Front Row, where most of such companies maintained business headquarters, counted among their clients men like Josiah Wilson of northeastern Arkansas. The Steele brothers managed Wilson's funds, including investments and the payment of his families' day-to-day expenses. When Wilson died in 1870, J. & J. Steele & Co. signed the bond for his widow when she petitioned the court to allow her to assume the responsibilities of executrix of her husband's estate. They continued to handle the family's investments and supplied the household with its necessaries, deducting the costs from the running account.[22]

E. Reed (Tuscaloosa and London: The University of Alabama Press, 1994), 97–120. Blacks constituted 15,442 out of a population of 40,226. For statistics on the Irish and African American population, see Census of Population, Vol. 1, Population and Social Statistics, 1870, Table 8, 380 and Table 31, 768.

22 Jeannie Whayne, *Delta Empire: Lee Wilson and the Transformation of Agriculture in the New South* (Baton Rouge: Louisiana State University Press, 2011), 57. A number of studies of the Memphis epidemic have been published but the most recent is Jeanette Keith's *Fever Season: The Story of a Terrifying Epidemic and the People who Saved a City* (City: Publisher, Year). Keith uses the biographies of several individuals who remained in Memphis to tell the story of the epidemic.

Even after the Memphis cotton elite formed the Memphis Cotton Exchange in 1873, the cotton factors continued contracting directly with planters and maintained the commission arrangement much longer than such men elsewhere. Their cotton exchange "formulated and enforced trading rules, provided its members with the latest information on cotton prices in key markets, arbitrated disputes among buyers, and sellers, and promoted Memphis cotton worldwide."[23] However, the Memphis Cotton Exchange engaged in no futures trading, unlike the New York Cotton Exchange, and served instead as a spot market. In other words, the exchange "made no provision for the buying and selling of contracts for the future delivery of cotton." In 1885 one St. Louis newspaperman opined that "the close relations that existed between planters and factors" in Memphis perpetuated the commission system.[24]

Some things changed more rapidly after the Civil War, however. A variety of cotton-related support industries also developed in Memphis; they included cotton oil mills and cotton seed companies, but large cotton compress firms were chief among the new concerns. "Memphis entrepreneurs built large warehouses to store the crop and powerful compresses to reduce the size of cotton bales." The expansion of the railroad system to Memphis played a major role in the manner in which the compresses operated. "Compressing gin bales to one-third their size made it possible to pack 18,500 pounds of cotton, or about thirty-seven bales, into a single freight car in 1873, creating a great saving in transportation cost." A new group of prosperous commission merchants began selling a variety of items to supply the needs of the rural inhabitants in the Memphis orbit. The increase in the acreage devoted to cotton cultivation made necessary the marketing of foodstuffs to feed the expanding rural population. Memphis commission operators not only supplied plantation commissaries, but also the many country stores and small town merchants who emerged in the post war period. In 1876 "the city's twenty-four leading firms were: eleven wholesale grocery and cotton commission businesses, two cotton factorage firms, seven wholesale dry goods companies, three wholesale boot and shoe companies, and a wholesale meat company."[25]

Despite this relative advantage, Memphis faced unprecedented problems, some of them with prewar origins. The city's leaders had convinced voters

23 Wrenn, *Crisis and Commission Government in Memphis,* 3.
24 Ibid., 3–5.
25 Ibid., 4–5.

to incur a heavy bonded indebtedness in the late 1850s, when cotton prices were high, in order to attend to its growing infrastructure problems. During the war the population expanded significantly, unfortunately at a time when the city could neither pay its debt nor address the deterioration of its roads and other services. The occupation of union troops made Memphis a mecca for freed people who abandoned the plantations on which they worked and flocked to the city to find safety within Union lines. The population continued to increase in the post war era as Memphis nearly doubled, from 22,263 in 1860 to 40,266 in 1870. This put additional pressure on the city's infrastructure at a time when tax revenues declined dramatically.[26] Given that the state of Tennessee had no meaningful tax enforcement measures, there was little city officials could do in the face of the unwillingness or inability of the city's property owners to pay the taxes necessary to meet the payments on the bonded indebtedness. Various schemes to scale back the debt were attempted after the war, but the city had little maneuvering room and bond holders were in no mood to compromise. When the state of Tennessee in 1875 passed a measure which would have allowed Memphis to pay at the rate of fifty cents on the dollar for its bonds, only a third of the bondholders agreed. The others took the matter to court. By early 1878, the city was in default and essentially bankrupt. "In 1880 Memphis had the highest per capita debt of any city in the South. Nationwide, Memphis would be remembered as 'the most notable municipal bankruptcy of the 1870s.'" Just when it seemed that things could not get worse, they did. A yellow fever epidemic which began later that year delivered a final blow to a city already over the edge economically.[27]

Ironically, the yellow fever epidemic gave the cotton elite an opportunity to recapture the reigns of city government and manage the city in a way that addressed their best interests.[28] Although no one could have anticipated such a development, the epidemic itself was hardly surprising. Throughout the antebellum period, towns along the lower Mississippi River valley had suffered periodic bouts of cholera and yellow fever, and they only increased in severity over time. Memphis experienced its first serious outbreak of the latter in 1828, just nine years after its founding, but the population was then very small, perhaps no more than 500 persons (according to the census of

26 This was true of other places in the Confederacy. Wrenn, *Crisis and Commission Government in Memphis, 20–21.*

27 Ibid., 16, 18–20.

28 Ibid., 24–25.

1830, the town included only 663 inhabitant). Of the 150 reported cases in 1828, only 53 were fatal. Because yellow fever has a tendency to strike more densely populated areas, as Memphis grew, its vulnerability increased. In 1850, its population had expanded to 8,843 and by 1860 it was the sixth largest city in the South at 22,623. An epidemic in 1855, then, probably struck a city that consisted of approximately 14,000 people, but Memphis narrowly averted a major outbreak. Only 220 people died.[29] The failure of the disease to appear between 1862 and 1866 was probably due to the fact that federal authorities could impose effective blockades on items and individuals moving into garrisoned cities like Memphis and New Orleans. This stands in contrast to expectations of certain Confederates. "When New Orleans fell to union forces in April 1862, most expected (and many hoped) that yellow fever would lay waste to the northerners." A Virginia newspaperman opined that New Orleans would prove to be "a prize which will cost them vastly more to keep than [it] is worth, if his Saffron Majesty shall make his usual annual visit."[30] But his Saffron Majesty did not visit New Orleans or Memphis.[31]

With the end of union occupation came an increased vulnerability to an epidemic, and such an event occurred within two years, striking Memphis in 1867, a year when the population probably numbered over 30,000 and taking the lives of 550 people, including the husband and four children of Mary Harris Jones ("Mother Jones"). Despite the tragedy this brought to the families who suffered losses because of the disease, the fatality figures were remarkably slight. That would change dramatically in the 1870s. In 1873, an estimated 2,000 people died, causing panic and hysteria but resulting in little in the way of preparation for dealing with another such occurrence. In fact, it is difficult to imagine how the city might have better prepared. Although

29 Whayne, *Delta Empire*, 40–41.

30 Quoted in J.R. McNeill, *Mosquito Empires: Ecology and War in the Greater Caribbean, 1620–1914* (New York: Cambridge University Press, 2010), 293. McNeill does not cover the situation in Memphis, but his analysis of the experience of New Orleans is particularly convincing and almost certainly applies to Memphis.

31 The incidence of malaria is harder to gauge as references to it are often characterized as fevers or intermittent chills and fevers. McNeil takes the position that malaria did not pose a great threat to the federal military, but Andrew Bell, in *Mosquito Soldiers*, argues otherwise, citing impressive statistics suggesting that both Confederate and Union regiments were frequently hit by the disease, particularly in the Arkansas region. Andrew McIlwaine Bell, *Mosquito Soldiers: Malaria, Yellow Fever, and the Course of the American Civil War* (Baton Rouge: Louisiana State University Press, 2010).

we now know how the disease is spread and have a vaccination that protects individuals from contracting it, in 1878 little was known about the causes of the disease and no vaccination existed. The yellow fever virus is most often transmitted by the female *Aedes Aegypti* mosquito and is classified as an acute viral hemorrhagic disease. In order to transmit the disease, the mosquito has to bite a person already infected and ingest a sufficient amount of the virus which then goes through a transformation in the mosquito's stomach. The mosquito can then transmit the disease through its saliva glands when it bites another person.[32]

By the summer of 1878, approximately 50,000 people lived in Memphis, and it is likely that the yellow fever virus was delivered on board one of the steamships that docked at the Memphis harbor or perhaps came in by a road that ran along the river from Mississippi. It had been known since July 27th that New Orleans was suffering an epidemic and the citizens of Memphis watched nervously as incidences of the disease slowly crept up river, striking Louisiana and Mississippi towns in an inevitable trajectory northward. On August 12th it surfaced in the north Mississippi town of Grenada, just 101 miles from Memphis, and the next day the first case in Memphis was confirmed. By August 14th panicked Memphians began to flee the city. Within a few days only about 20,000 people, about 14,000 blacks and 6,000 whites, remained in the stricken city. By the end of the epidemic, 946 blacks and 4,204 whites had died. That represents a death toll of 6.0 percent for blacks and of 70 percent for whites. Although African Americans were not immune to the disease, they did enjoy a resistance that made it less likely to be fatal to them. This had been long understood and largely accounts for the

32 Bell, *Mosquito Soldiers*, 28–29; Todd L. Savitt, *Medicine and Slavery: The Diseases and Health Care of Blacks in Antebellum Virginia* (Urbana: University of Illinois Press, 1978), 18–19, 25; Margaret Humphreys, *Yellow Fever and the South* (Baltimore: The Johns Hopkins University Press, 1992); Humphreys, *Malaria: Poverty, Race, and Public Health in the United* States (Baltimore: Johns Hopkins University Press, 2001); John H. Ellis, *Yellow Fever and Public Health in the New South* (Lexington: University Press of Kentucky, 1992), 27–28; John R. Pierce and Jim Writer, *Yellow Jack: How Yellow Fever Ravaged America and Walter Reed Discovered its Deadly Secrets* (New York: J. Wiley, 2005); Molly Caldwell Crosby, *The American Plague: the Untold Story of Yellow Fever, the Epidemic That Shaped Our History* (New York: Penguin, 2007); See also Charles Davis and William C. Shiel, Jr., "Malaria," http://www.medicinenet.com/malaria/article.htm (accessed June 19, 2011); and Mary T. Busowski and Burke A. Cunha, "Yellow Fever," http://emedicine.medscape.com/article/232244 (accessed June 19, 2011). See also S. Wright Kennedy, "The Yellow Fever Epidemic in Memphis, Tennessee: An Historical Geographic Information System (HGIS) Approach", MA Thesis (Long Beach State University, 2011).

decision of much of the black population to remain in Memphis during the epidemic. It also had certain short term political implications for them as African Americans played a crucial role, as either militiamen or police officers, in maintaining order in the city. This was not lost on the city's white elite.[33]

Among those who fled the city were many of the principal merchants, businessmen, and bankers, who set up operations in St. Louis. They included the organizers of the Memphis Cotton Exchange, founded in 1873. Its principal purpose had been to promote the sale of the city's cotton on the world market and, particularly, to maintain the important relationship with buyers in Liverpool.[34] A few cotton brokers and prominent businessmen remained in place, however, and survived the epidemic. They would play a crucial role in positioning the cotton elite to use the yellow fever crisis as an opportunity to lobby the governor and seize control of the municipal government.

Given the disarray arising out of the epidemics, these businessmen successfully represented themselves as speaking for the city and were permitted to voluntarily surrender the city's charter in 1879. The city became the "Taxing District of Shelby County" with a director and two of four commissioners appointed by the governor. The cotton elite dominated the appointments and at their behest, the legislature voted to repudiate Memphis' debt. However, after eastern financial institutions raised a hue and cry, leaders in the Memphis Cotton Exchange worked with the state to establish a reasonable alternative, one which would honor the debt and allow the city to recover even as it paid a small portion on the debt for a period of years. Meanwhile, they dismantled the ward system and substituted the new commission form of government, one that called for city-wide elections.[35] The Irish and African-American coalition lay in shreds. Their political ascendancy had depended on the first, ninth, and tenth wards but "from 1879 through 1893, not a single commissioner lived in the first and ninth wards of North Memphis nor in the tenth ward of South Memphis."[36] Yet the cotton elite did not move to disfranchisement and, instead, capitalized on the black vote to maintain their control until the early 1890s. The state, working with the city's leader-

33 Whayne, *Delta Empire*, 41; Capers, *The Biography of a River Town*, 189; Keith, *Fever Season*, 144; Wrenn, *Crisis and Commission Government*, 23.

34 Janie V. Paine, *The Memphis Cotton Exchange: One Hundred Years* (Memphis, The Memphis Cotton Exchange, 1973), 4.

35 Wrenn, *Crisis and Commission Government in Memphis*, 33–33, 94; Keith, *Fever Season*, 78–79.

36 Ibid, 143.

ship, began the process of addressing some of the serious and long-standing problems confronting Memphis, particularly the lack of a sewer system and a clean water supply.

The period between 1860 and 1880 had been a tumultuous one which witnessed a series of challenges, from military occupation and interruption in the cotton trade during the Civil War to economic chaos, unprecedented political challenges, and major outbreaks of yellow fever after the war. An important aspect of the recovery of the cotton economy was the ability of white land holders to manipulate the black population and impose the sharecropping system on the freed people. Although often referred to as a "compromise," the end result—debt peonage and impoverishment of the freed people—rendered that compromise a bad bargain.[37] For landowners, however, the bargain worked a kind of magic, returning them to financial supremacy. By 1880, cotton production had surpassed pre-war levels, rising by over 75 million pounds, from 231,139,500 pounds in 1859 to 306,660,000 pounds in 1879. Although the old business elite in Memphis had recaptured control of the city and its finances, they were to face another series of problems in the decades to come.

1880–1900

In the period between 1880 and 1900, even as Memphis struggled to recover from the yellow fever epidemic and the city's longstanding financial woes, it enjoyed a revival of the cotton economy. Although cotton prices continued to be lackluster, the acres in production and the number of cotton bales produced increased significantly. It was during this twenty-year period that the Mississippi and Tennessee cotton producers recovered their footing and

37 The literature on the sharecropping and tenancy system is extensive but must begin with Pete Daniel's *Shadow of Slavery,* (City: Publisher, Year), a crucial book that established the field. Also contributing crucial insights are the following: Harold D. Woodman, *New South, New Law: The Legal Foundations of Credit and Labor Relations in the Postbellum Agricultural South* (Baton Rouge: Louisiana State University Press, 1995); Roger L. Ransom and Richard Sutch, *One Kind of Freedom: The Economic Consequences of Emancipation* (Cambridge, UK and New York: Cambridge University Press, 1977); Leon Litwack, *Been in a Storm so Long: The Aftermath of Slavery* (New York: Knopf, 1979); and Douglas A. Blackmon, *Slavery by Another Name: The Re-enslavement of Black People in America from the Civil War to World War II* (New York: Doubleday, 2008).

Arkansas growers increased the volume of production. The Tennessee growers, however, began to turn to other crops, perhaps because they had exhausted their cotton lands. Although the amount of acres in improved farms increased significantly, the cotton bales harvested in the Tennessee counties decreased. The rich delta lands in Mississippi and Arkansas, on the other hand, were worth developing and much of the newly reclaimed acreage was planted in cotton. Between 1880 and 1900, acreage in farms in the Memphis orbit rose by another 1,632,575 acres while cotton production figures rose by nearly 60 million pounds, from 306,660,000 to 366,079,500.

The elite's new political ascendancy, accomplished after overturning the ward system in favor of commission government, had freed them to fight amongst themselves over a number of issues, particularly how to fund improvements to the city's streets, school financing, and debt settlement. One issue they had little disagreement over, however, involved an environmental problem that threatened them all.[38] Settlers along the upper Mississippi River and its tributaries had cleared and drained millions of acres of land; in doing so they redirected an enormous amount of water and debris into the fast-flowing river. By the time it reached the Memphis area, the river was a virtual monster, unpredictable in both flow and ferocity. It inundated farmland just as new settlers were flocking into the region and demanding attention from a federal government unaccustomed and largely unwilling to address unprecedented demands that farm lands be protected from floods by the national government. Competing ideas about how to control the river and keep its channel open to commerce complicated the government's response. Some advocated a line of levees that channeled the river's flow in a predictable way. Others supported the use of levees together with creation of a series of controlled watersheds meant to allow the river to expand into certain reservoirs whenever necessary. The "levees only" approach won the debate.[39]

In 1876 Congress authorized the creation of the Memphis District of the Corps of Engineers, a response to the growing threat of floods to Memphis and the surrounding area, and in 1879 Congress established the Mississippi

38 Wrenn, *Crisis and Commission Government in Memphis*, 146–147.

39 Ann Vileisis, *Discovering the Unknown Landscape: A History of America's Wetlands* (Washington, D.C.: Island Press, 1997), 71–73. Andrew A. Humphres, Joseph B. Eads, and Charles Ellet, Jr. were the three most prominent and influential engineers operating in the nineteenth century, see John M. Barry, *Rising Tide: The Great Mississippi Flood of 1927 and How it Changed America* (New York: Simon & Schuster, 1997).

River Commission (MRC). Although the latter was headquartered in St. Louis, it maintained an office in Memphis and held frequent meetings there to address the specific needs of the city's residents and those of farmers and planters in the region. "With this, the federal government began to erect the bureaucratic infrastructure necessary for those interested in bringing the river under man's control. Composed of seven presidential appointees, three of them officers in the Corps of Engineers, the MRC's chief responsibility included 'the task of preparing surveys, examinations and investigations to improve the river channel.'" Appropriations were tied to improvements in the river's navigability and, importantly for Memphis, maintaining the nation's harbors in order to facilitate commerce and trade.[40] A particularly serious flood in 1882 led Congress to pass the Rivers and Harbors Act which formalized the relationship between the Corps of Engineers and the MRC. The MRC was to serve as the planning agency and the Corps of Engineers was to assume responsibility for carrying out improvements to rivers and harbors.[41]

Memphis businessmen and cotton brokers welcomed the greater commitment on the part of the federal government and hoped that soon federal funds would aid them in repairing the damage to the city's harbor. Memphis was better connected by railroad lines by this time, but the river remained the most important component of its commercial success, receiving and dispatching tens of millions of pounds of cotton annually, among many other business-related necessities. The flood of 1882 had damaged the Memphis harbor and another flood in 1883 caused further destruction, so Congress appropriated $200,000 in July 1884, to remedy the situation. The problem with the Memphis harbor, however, was complex and required a multifaceted approach. The most obvious issue was that the flooding had overwhelmed and inundated the harbor, damaging its structures and facilities. While buildings and docking facilities could be rebuilt, two more fundamental problems existed, one involving the city's custom of draining its increasing flow of waste water onto the river's bluff. This caused landslides of the easily fractured Chickasaw Bluff. It was solved by enclosing the city drains in wooden culverts and diverting the waste water from the bluff down to the river's edge.[42]

40 Whayne, *Delta Empire*, 64. See also Floyd M. Clay, *A Century on the Mississippi: A History of the Memphis District, U.S. Army Corps of Engineers, 1876–1981* (U.S. Army Corps of Engineers, Memphis District, 1986), 14–17.

41 Clay, *A Century on the Mississippi*, 33–36.

42 Ibid., 33–36.

The second problem was more complicated and was caused by a peculiar set of environmental issues. The river had been damaging the bluff, eating away at it relentlessly, and by 1874 engineers estimated that a hundred feet per year along the water front was being destroyed. Meanwhile, an eddy current developed along the Memphis harbor, making it difficult for steamships and small craft alike to put into port there. It also caused flooding onto Market Street, which fronted the river, and threatened to destroy it. An eddy current typically develops on the downstream side of an obstruction protruding into a river, but this one resulted from the recession of Hopefield Point, on the Arkansas side of the river, a particularly unusual occurrence. Hopefield Point essentially protruded into the river which slammed into the point and tore off "great chunks" of it and deposited "heavy loads of silt before the city." That would lead eventually to the formation of a bar in front of the Memphis harbor, which became Mud Island twenty years later, but in the 1880s an eddy current developed as the recession of Hopefield Point also caused a "counter current running from the length of Market Street to the mouth of the Wolf River." As the point's recession rate increased, the eddy current grew more severe. Eddy currents are especially treacherous in fast-moving rivers carrying debris such as the Mississippi, and the one that confronted Memphis manifested as a giant whirlpool or a swirling maelstrom of water embedded with trees and branches that acted as projectiles dangerous to steamships putting into port.[43]

The corps began to lay plans to address the problem, but the MRC, which had charge of the Corps of Engineers, first thought it necessary to study the situation. The engineers argued over an appropriate remedy, and the entire matter became embroiled in city, MRC, and Congressional politics. One particularly contentious argument arose over how to solve the problem involved with revetment of the banks along the city's waterfront. One engineer recommended that "concrete be poured in sections, and that those sections be connected to each other by iron rods." As the author of a book on the Memphis District wrote, it was a "mature concept" but at the time "there was neither the money nor the technology for expensive experimentation."[44] The use of concrete mattresses would become commonplace in the 1930s but in the 1880s the Corps still relied on the use of willow mattresses. The technique, developed in Germany in the twelfth century, involved placing

43 Clay, *A Century on the Mississippi,* 24, 34–35.
44 Ibid., 33–34.

the mattress as a kind of "protective cloak on the bank itself" to keep it from eroding into the river. The mattresses were constructed by cross-stacking willow brush in three- to five-foot layers which were tied together. Once constructed, they were floated into position and secured to the bank with rocks. They worked pretty well on most rivers but would often prove to be inadequate along much of the fast-moving, debris-laden Mississippi River.[45]

The Corps ultimately decided to use the willow mattresses to solve the problem at Memphis but as late as 1886, no plan was yet in place and the eddy current had virtually destroyed 900 feet of the lower river front. It was only then that the Corps, acting in concert with the city, determined upon a remedy. It would build "spur-dikes out from the landing areas in order to break up the current." The Corps had designed the five dykes, which were "constructed by building a wooden crib, floating it into position, and then sinking and anchoring it in place by filling it with stone." They fastened the dykes to the shore on willow mattresses constructed for the purpose. This was hardly going to stand the test of time but it solved the problem temporarily. "The dykes not only broke up the current which had been eroding the banks, they also provided anchorage for the steamboats." Because the city contributed most of the $60,000 necessary for their construction (with the Corps paying the balance), they became known as "'citizen dikes," and the Corps used this strategy—of building dykes and sharing the costs—with other cities facing similar problems.[46]

The cotton elite and the city's other business leaders were united in their support of improvements to the city harbor, but this was an uneasy alliance in other respects. The cotton elite maintained its control of municipal government in part by appealing to the black vote. Though both black and Irish politicians suffered a loss of electoral office during this period, they continued to deliver voters and their respective communities received some benefits. As disfranchisement efforts swept across the South, Democrats in favor of disfranchisement swept into office in 1888 in Tennessee and enacted disfranchisement measures aimed mostly at the cities of Memphis, Nashville, Chattanooga, and Knoxville. In Memphis, those who resented the cotton elite's control welcomed the opportunity to undermine them by eliminating one major source of their power—the black vote—and successfully unseated the cotton elite in municipal elections in 1890. By the time Memphis sur-

---

45 Clay, *A Century on the Mississippi,* 31.
46 Ibid., 33–36.

rendered the taxing district arrangement and reclaimed its city charter in 1893, the cotton elite no longer remained in control over city government, but they remained a preeminent economic force and thus maintained some political influence.[47]

Despite two decades of environmental and political turmoil, Memphis was able to return to solvency and emerged from this period stronger than ever. The rededication of planters in the countryside to the cultivation of cotton worked to the benefit of the city's businessmen. Even though the country and the region suffered from a serious recession in the 1890s and cotton prices bottomed out, lowering to a nickel a pound in 1894, a recovery had stabilized by 1900. Cotton prices had returned to pre-recession levels, reaching nine cents a pound in 1900, a sum that would have resulted in a return of $32,947,155 to growers in the region. The cotton factors, commission merchants, bankers, and businessmen of Memphis likely received at least half of that for their services to the community of planters and farmers.

1900 to 1920

As it stood in 1900, Memphis was thriving, and its businessmen were happy. Cotton factors could look forward to an ever increasing flow of cotton as production had reached new heights and was destined to increase by 95 million pounds over the next twenty years. By 1920, even Tennessee had rededicated itself to cotton, though its yields were no better than they had been twenty years earlier. Mississippi and Arkansas growers had increased their acreage devoted to cotton and the number of bales they produced but, like Tennessee, their yields were declining. High prices during World War I probably account for much of their rededication to cotton, but their yields suggest they were likely bringing into cultivation marginal lands. Most of the new acres in cultivation came from Arkansas, but even Tennessee planters increased their improved acres. Even though the Arkansas growers operated under some special environmental handicaps, they were producing more cotton than the east Tennessee planters. What is remarkable about this "achievement" is that it was during this period that Mississippi River flood stages reached record levels, swamping farmlands and endangering the

47 Wrenn, *Crisis and Commission Government in Memphis*, 143–144.

Memphis harbor. To complicate the situation, America's involvement in the Great War led to record-breaking cotton prices but also diverted the attention of the Corp of Engineers away from the problems along the lower Mississippi River valley. By the time the war was over and the agency shifted its attention to the difficulties facing planters in the valley, a postwar recession struck and congressional appropriations reflected that reality.[48]

In a sense, this was the best of times and the worst of times for the Memphis cotton economy. Farmers across the country enjoyed greater prosperity in the period between 1900 and 1920 than had known in decades. This was as true for the cotton planters in the Memphis orbit as it was for those elsewhere, and the city benefitted from this development. But Memphis was also undergoing some significant political and economic changes that would, over the long term, reduce the importance of cotton. Although the staple remained dominant for the time being, new economic forces were taking shape. For example, Memphis had become the largest hard-wood lumber center in the country as entrepreneurs, some southern born, some from elsewhere, had established lumber companies throughout the region, shipping millions of board feet of lumber to Memphis yearly to be sold nationally and internationally. Meanwhile, New South boosters numbered among the Memphis elite and were intent on promoting industrial development. They found support from E.H. "Boss" Crump who served as mayor from 1910 to 1915 but then became a political power-broker behind the scenes for almost four decades. He astutely maneuvered both the black vote and that of white Republicans to maintain control. Blacks in Memphis, unlike those in other southern cities where blacks were largely disfranchised, remained viable political players. No one living through the first two decades of the twentieth century could have missed the importance of Crump's ascendancy, but, given the prosperity in the farm economy, particularly during World War I, no one could have grasped the challenge to the cotton elite that Crump's long-term plan for the city represented.[49]

48 Cotton prices began the decade at 9.65 cents per pound in 1911 and rose accordingly: 1912, 11.50 cents; 1913, 12.47 cents; in 1914 the price dropped to 7.35 cents as a result of the outbreak of hostilities but then began to rise again: in 1915, 11.22 cents; and in 1916, 17.36. The bonanza began in 1917 and lasted through 1919: 27.09 cents, 28.88 cents, and 35.34 cents. United States, Historical Statistics of the United States: Colonial Times to 1970. Washington: U.S. Dept. of Commerce, Bureau of the Census, 1975. Tables 550–563. Pp. 517–518.

49 William D. Miller, *Mr. Crump of Memphis* (Baton Rouge: Louisiana State University Press, 1964); G. Wayne Dowdy, *Mayor Crump Don't Like It: Machine Politics in Memphis* (Jack-

Indeed, the challenges facing the Memphis elite seemed to be not of man's making at all. While river stages had been rising during the second half of the nineteenth century, particularly in the 1880s and 1890s, the situation worsened in the early twentieth century.

River Stages, Memphis[50]	
1874	34.00 feet
1882	35.15 feet
1898	37.22 feet
1903	40.10 feet
1912	45.23 feet
1913	46.55 feet

Record-breaking high water in 1903 rose to a flood stage of 40.1 feet at Memphis and lasted 94 days. While the levee held along the city's riverfront, the flood waters backed up into both the Wolf River and Bayou Gayosa and flooded portions of Memphis anyway. Hundreds of thousands of acres were inundated in the surrounding countryside, particularly in Crittenden and Lee counties in Arkansas. There the levees failed miserably. While Memphis had survived the flood with relatively minimal damage, a second flood in 1907 caused more destruction to the city than anywhere else in the region. A section of the bluff collapsed "with a thunderous roar," swamping steamboats as much as a half mile away on the river. To make matters more bizarre, "north Memphis flooded and 2,000 barrels of mash were set afloat in a distillery." But as much trouble as the 1903 and 1907 floods caused in Memphis and along the river valley, they hardly compared to the devastation wrought by the floods of 1912 and 1913. The 1912 flood took out fourteen miles of levees and flooded 10,812 square miles of land. The 1913 flood was even more severe. Fully 24 percent of the area covered by the Memphis District of the Corps of Engineers was under water. For two years before the disasters, Congress had been allotting less money for rivers and harbors. President William Howard Taft believed that railroads were carrying more of

son, Mississippi: University Press of Mississippi, 2006); and Roger Biles, *Memphis in the Great Depression* (Knoxville: University of Tennessee Press, 1986).

50 Mississippi River Commission, *Annual Highest and Lowest Stages of the Mississippi River and its Outlets and Tributaries to 1960* (Vicksburg, MS: U.S. Army Engineer District, 1960), 54.

the nation's freight and that rivers were no longer so crucial to moving cargo and insuring the flow of commerce. He made this announcement in 1910 at a Rivers and Harbors Convention, and for two years Congress appropriated smaller sums for river revetment. The devastation wrought by the 1912 and 1913 floods made that policy seem short sighted, and soon appropriations returned to their former levels. However, the Great War diverted the attention of Congress and the Corps of Engineers away from domestic concerns, and smaller sums were expended on keeping the river channels open and the harbors secure.

The 1913 flood also had the effect of completing another task the river had set about accomplishing. For several decades a new bar—a mass of sand and silt -- had been rising in front of the city's harbor. Together the floods and the increased flow had worked relentlessly against the old Hopefield Point recession. Just as the river had created the recession that led to the eddy current that threatened Memphis in the 1880s, it also undermined and further destroyed the recession in the 1890s. As the recession dissolved, its silt and sand drifted straight toward Memphis, playing the major role in creating the bar. Two secondary factors contributed: sawdust from mills operating along the Wolf River found its way into the harbor (the Corps estimated 20 percent of the debris coming out of the Wolf River was sawdust) and much of it gravitated toward the rising bar, and the collapse of a segment of the bluff, which slid straight into the bar, during the 1907 flood. As it happens, however, the bar had the potential to protect the harbor and its structures from the onslaught of the river's current. At the urging of certain Memphis citizens and businessmen, the Corps dredged the channel between the bar and the harbor and riveted the bar—which became known as Mud Island—to make it a permanent feature. In the end, the river's current had worked to the advantage of the city.[51]

By the time that the war was over and the Memphis District of the Corps was positioning itself to return to its local responsibilities, a serious postwar recession struck and Congress restricted funding to pre 1912/1913 levels. By this time, the Corps was ready to assume responsibility for flood control and had a better understanding of how to accomplish the task before it, but they had no mandate from Congress and, more importantly, insufficient funding. The Mississippi River Valley was ill prepared for the fate awaiting it in the next decade as river levels continued to rise, leading ultimately to the disas-

51 Clay, *A Century on the Mississippi*, 54, 59–60.

trous flood of 1927 and another—and more permanent—reversal of policy on the part of Congress.[52]

Throughout this period cotton production had been on the rise and fortunes were made, both for the planters and farmers in the city's orbit and for the cotton elite in Memphis. Cotton prices tell the tale:

Year	*Cents per Pound*
1899*	7.0
1900	9.2
1901	7.0
1902	7.6
1903	10.5
1904	9.0
1905	10.8
1906	9.6
1907	10.4
1908	9.0
1909	13.5
1910	14.0
1911	9.7
1912	11.5
1913	12.5
1914	7.4
1915	11.2
1916	17.4
1917	27.1
1918	28.9
1919*	35.3

To put it another way, the farmers in the Memphis orbit reported harvesting 366,079,500 pounds worth $25,625,565 for the 1899 crop year. In 1920, they reported 460,804,000 pounds harvested worth $161,663,812 for the

52 Clay, *A Century on the Mississippi*, 54, 56, 61.

1919 crop year.[53] These were high times indeed, for farmers and those who did business with them.

However, just as the floods undermined the Memphis harbor, they also challenged farmers and planters in the city's orbit who began to call on the Corps to address the problem. The Corps of Engineers and the MRC had only a limited mandate from Congress: keep the harbors functioning and river channels open in order to facilitate trade. Thus the Corps only constructed levees alongside farmlands when they could make the argument that they helped to stabilize the river channel. To meet the necessity of providing protection against floods, the states of Tennessee, Mississippi, and Arkansas stepped into the breach and began to take action. For example, the state of Arkansas created the St. Francis Levee District in 1893 and endowed the district with responsibility for flood control in eight counties—Mississippi, Crittenden, Craighead, Poinsett, Cross, St. Francis, Lee, and Phillips—encompassing 2,500 square miles, or 1.6 million acres. The district comprised some of the most fertile land in the United States, but much of it was in swamps or subject to periodic overflows. The St. Francis Levee District began to construct levees to protect the farmers and planters in the district against the annual overflows. Meanwhile, Arkansas counties, like those elsewhere in the Memphis orbit, had been building some levees to protect its towns and farmlands, but in the early twentieth century also began to create drainage districts in order to drain the swamps, an endeavor highly favored by the planters and most of the farmers in the lower Mississippi River Valley.[54]

The creation of county drainage districts in Arkansas was made possible by a law passed by the state legislature in 1902 allowing counties to sell bonds to fund the construction of canals to drain overflowed lands. Landowners wishing to create a drainage district were required to petition the county court for permission to do so, the county court was required to review the petition in open court, and landowners could then voice their support or

53 Pounds harvested in 1899 are reported on the 1900 census just as pounds harvested in 1919 are reported on the 1920 census. Because my figures are drawn from the 1900 and 1920 censuses, I used the price per pound in 1899 when figuring the value of the crop reported in 1900 and the price per pound in 1919 when figuring the value of the crop reported in 1920. The price per pound was derived from *Historical Statistics of the United States, Earliest Times to the Present*, Millennial Edition, Vol. 4, Crops and Livestock, Series Da755–765, 4–111 (Cambridge and New York: Cambridge University Press, 2006).

54 St. Francis Levee District, *A History of the Organization and Operations of the Board of Directors, St. Francis Levee District of Arkansas, 1893–1945* (West Memphis: St. Francis Levee District, 1945), 162–67, 179–80.

concerns. Some landowners opposed the creation of districts because they were unwilling to bear the tax burden they would incur, but proponents argued that the benefits outweighed the taxes and, after all, the taxes were necessary in order to repay the bond holders. The experience in Mississippi County, Arkansas, which is located in the northeast corner of that state, stands as an example of the efficacy of drainage districts in terms of cotton production. In the first two decades of the twentieth century, the county created seven drainage districts (and several smaller sub districts) and engineers constructed 1,034 miles of ditches. The acreage in farms rose from 126,684 to 277,670 and cotton production from 34,380 to 112,778 acres. In other words, while the acres in farms more than doubled, the acres in cotton tripled.[55]

Most of the cotton grown in Mississippi County was destined for Memphis and, at the same time, Memphis played a more than incidental role in the creation of the drainage districts in the county. Memphis engineers, Arthur E. Morgan and Leroy L. Hidinger, for example, designed many of those drainage districts, and although they hired local crews to build the canals and ditches, most of the materials and machinery came from the city.[56] Technology, then, also played a role. For example, cross-walkers and draglines, transported first by railroad and then by oxen to remote locations, were instrumental in the construction of the ditches. A steam driven cross-walker, which looked like a house on a bridge with a small crane on it, spanned the ditch it was creating. It "walked" along by means of a moving shoe on both ends and excavated as it went. Drag lines were also used to excavate ditches, and dredge boats, sometimes constructed onsite, often finished out the project or were used to remove obstructions that later developed. Drainage would not have been possible without this technology and without the expertise of the engineers and those who ran the machinery.[57]

Even as the planters in the Memphis orbit began to take the steps necessary to protect their acreage from floods and to drain their swamplands, they faced another long-standing challenge. The problem of disease along

55 *Twelfth Census of the United States*, Agriculture, Part I, Vol. 5, 1900, (Washington: United States Census Office, 1902), 430 [for crops grown in 1899]; *Fourteenth Census of the United States*, Agriculture, Part 2, Vol. 6, 1920 (Washington: Government Printing Office, 1922), 479 [for crops grown in 1919]. For the controversy over the creation of drainage districts in Mississippi County, see Whayne, *Delta Empire*, 70–75.

56 Elliott B. Sartain, *It Didn't Just Happen* (Osceola, AR: Grassy Lake and Tryonza Drainage District NO. 9, 197?), 6.

57 Ibid., 16–21.

the Mississippi River Valley in this period continued to bring a different kind of disaster. While yellow fever began to decrease in severity as a result of scientific breakthroughs, malaria had become epidemic and a deadly variety reached the valley.

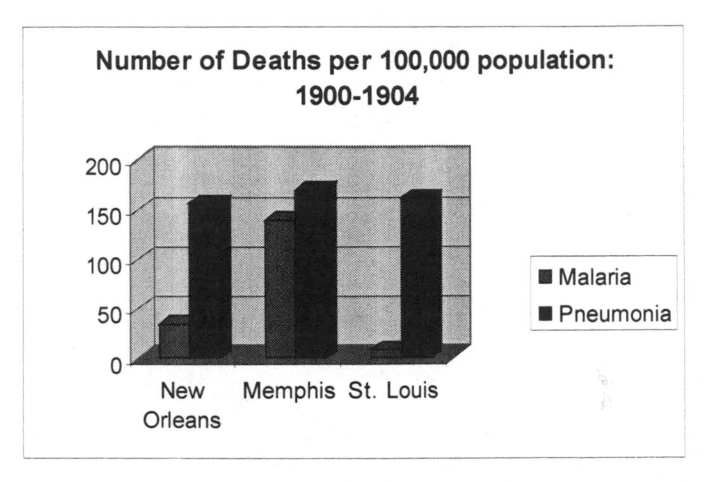

Drawn from Department of Commerce and labor, Bureau of the Census, Special Reports, Mortality Statistics, 1900 to 1904, Washington, D.C: Government Printing Office, 1906.

Significantly, the lower Mississippi River Valley constituted one of the worst malarial zones in the country. In 1906 the United States Bureau of the Census published mortality figures that demonstrated Memphis' vulnerability.[58] Of all the cities in the survey, Memphis was one of the unhealthiest. In fact, it had the highest percentage of deaths due to malaria of any city in the survey. Only pneumonia and tuberculosis outdid malaria as a cause of death in Memphis.[59] Indeed, as late as 1938, the lower Mississippi River Valley area from Cairo, Illinois, to Natchez, Mississippi, was identified as one of the four worst malarial zones in the United States.[60]

58 Department of Commerce and Labor, Bureau of the Census, Special Reports, *Mortality Statistics, 1900 to 1904* (Washington, D.C: Government Printing Office, 1906).

59 Ibid., 44–45.

60 American Association for the Advance of Science, Bulletin 15, "A Symposium on Human Malaria with Special Reference to North America and the Caribbean Region" (Washington, D.C.: Smithsonian Institution, 1941), 10. The relevant section reads as follows: "Heavily endemic foci [of malaria] that have remained relatively constant during this decade [1929–1938] are found in (1) a wedge-shaped sector of Southeastern States, including

Complicating the situation for the Memphis region was the arrival of the deadly plasmodium falciparum. The most common of the four types of the disease that typically struck humans, plasmodium vivax, sometimes caused death in very young children and could cause death in individuals with already compromised immune systems but the latter was rare. Plasmodium falciparum, however, appeared in the region in the late nineteenth century and could lead to death in an otherwise healthy adult. Some medical professionals had begun to suspect that the mosquito played a role in the spread of malaria but had no idea that an ancient African parasite was the culprit. In fact, the parasite is transmitted by the female *Anopheles*. When it bites an infected person, a transformation can occur within the mosquito which makes the parasite transferable, and sometimes lethal, to the next person the mosquito bites. [61]

While an endemic disease like malaria does not present the same kinds of economic challenges as do floods and other disasters, it introduces instability, saps vitality, and erodes confidence. Those who study the cost of disease refer to the "disease burden" and calculate the reduced quality of life, the loss of income from a life cut short, and the financial costs to the community, among other things. For cotton planters and farmers, the costs would be directly felt in terms of loss of productive labor, particularly at the harvest but, in fact, at every important stage along the crop cycle.[62]

Given the environmental issues confronting the region in this period, the growth in the cotton economy is all the more remarkable. More than three-quarters of a million new acres had been brought into cultivation, contributing to the production of an additional 94,724,500 pounds of cotton between the 1900 and 1920. A drop in cotton prices in the early 1920s would lead to massive foreclosures in the agricultural sector over the next few years and the bankruptcy of cotton factors and businessmen in Memphis. Indeed, the cotton factorage industry was hit with a devastating crisis from which it never fully recovered. For example, "the number of cotton factors operating

areas in South Carolina, Georgia, Florida, and two adjacent counties in Alabama; (2) the 'delta area' of the lower Mississippi Valley from Cairo, Illinois, to Natchez, Mississippi; (3) the portion of the Red River Valley near the junctions of Oklahoma, Arkansas, Louisiana and Texas; and (4) a 4-county section of Texas at the mouth of the Rio Grande."

61 Whayne, *Delta Empire*, 22.

62 For "disease burden," see The World Health Organization, http://www.who.int/topics/global_burden_of_disease/en/.

in eastern Arkansas dropping from sixty-four to two" by 1921.[63] Because of the intimate connection of the city's businessmen to the farm economy in the region, Memphis would find the next decades to be rough going.

Conclusion

During the period between 1840 and 1920, Memphis evolved into a great river city, a cotton metropolis. However, the dedication to cotton processing and marketing to the exclusion of other economic enterprises provided only a limited economic foundation.

Cotton production initially rested on the slave labor system which retarded economic development within the South and in the Memphis region. Most of the capital was tied up in land and slaves with little left for investment in either infrastructure or industrial development. The implications were as true for the hinterlands as it was for the city. Cotton cultivation required only men, mules, and plows, and even the processing phase necessitated only the cotton gin. Most of the processed cotton was sent to England through cities like Memphis—with only a small percentage delivered to the newly-developing textile industry in the northern U.S. to be spun into cloth. When the slave labor system gave way to sharecropping, another form of unfree labor, after the Civil War, the South rededicated itself to the labor intensive plantation system and the production of cotton while farmers in the Midwest took advantage of machine technology, mechanized, and purchased reapers marketed by a growing number of Chicago firms. This study of Memphis and its relationship to the expansion of cotton cultivation in the adjacent countryside draws inspiration from William Cronon's *Nature's Metropolis*. Like Chicago, Memphis was situated in the perfect location to serve as an important entrepot, but Memphis functioned in a political economy utterly unlike the one that propelled Chicago to greater heights.[64]

Sitting on the western edge of the cotton kingdom, Memphis would see cotton production expand by nearly 4 million new acres and 320 mil-

63 The World Health Organization, http://www.who.int/topics/global_burden_of_disease/en/. 102; and B.M. Gile, "Organization and Management of Agricultural Credit Associations," *Arkansas Agricultural Experiment Station Bulletin* 259 (1931), 31.

64 William Cronon, *Nature's Metropolis: Chicago and the Great West* (New York and London: W.W. Norton & Company, 1991).

lion pounds in the fifty years after the Civil War. This growth would inspire railroad promoters who extended lines into Memphis, but the Mississippi River remained its most important conduit, particularly during the nineteenth century. However, the river was a fickle and unpredictable ally. Its annual overflows inundated farmlands, threatened small communities, and damaged the Memphis harbor thus interrupting the flow of cotton and commerce. The river only grew fiercer in the late nineteenth century as the upstream activities of settlers along the Mississippi and its tributaries resulted in an increased flow of water, silt, and debris and led the planters and farmers in the hinterlands and the businessmen and cotton factors in Memphis to launch efforts to bring the river under control. They would use every means at their disposal to remedy the situation, including their influence with local, state, and federal governments. Once commitments had been made to attempt to bring the river under their control, a body of government bureaucrats and engineers combined to consolidate power in the cotton elite, both those in Memphis and in its hinterlands.[65] Under these circumstances the businessmen and cotton factors of Memphis grew wealthy, and the cotton economy, for good and ill, fastened its grip on the region. A cotton metropolis had emerged and stood at the nexus of the region's economic development.

65 Vileisis, *Discovering the Unknown Landscape*, 71–73. Historians familiar with Donald Worster's analysis in *Rivers of Empire: Water, Aridity, and the Growth of the American West* (Oxford: Oxford University Press, 1992), would find unsurprising the coalescence that occurred in the late nineteenth century.

TABLE I
IMPROVED ACRES IN FARMS, 1850 TO 1820

	1850	1860	1870	1880	1900	1920
Arkansas	65,501	175,426	123,783	286,179	722,311	1,001,979
Mississippi	726,319	1,080,448	949,670	1,193,344	1,757,181	2,529,951
Tennessee	1,031,449	1,178,936	1,389,478	1,794,471	2,427,259	2,653,307
Total	1,823,269	2,434,810	2,462,931	3,273,994	4,906,751	6,185,237

[Bureau of the Census, *Eighth Census of Agriculture of the United States,* 1860, Vol. III (Government Printing Office: Washington, 1864), pp. 8–9, 84–85, and 132–133; *Ninth Census of Agriculture of the United States*, 1870, Wealth and Industry of the U.S. (GPO, 1872), pp. pp. 100, 184, 244, 246; *Tenth Census of Agriculture*, June 1, 1880, Part I (GPO, 1883), pp. 105, 122–1234, 133; *Twelfth Census of Agriculture*, 1900, Vol. 5, Pt. 2, Southern States (GPO, 1902), pp. 167, 168, 284, 285, 297, 298;and *Fourteenth Census of Agriculture*, Vol. 5, Pt. 2, Southern States, 1920 (Washington: GPO, 1922), pp. 446–454, 464–472, 522–529, 538–545, 560–567, 575–581.]

TABLE II
POPULATION, 1850 TO 1920

	1850	1860	1870	1880	1900	1920
ARKANSAS						
Slaves/Blacks	5,243	13,835	18,007	40,675	73,492	101,585
Free/Whites	13,495	13,482	13,295	26,236	38,614	58,460
Total	**18,738**	**27,317**	**31,302**	**66,911**	**112,106**	**160,045**
MISSISSIPPI						
Slaves/Blacks	63,307	95,346	105,260	174,588	237,207	293,440
Free/Whites	59,102	58,336	75,690	88,066	99,719	107,395
Total	**122,409**	**153,682**	**180,950**	**262,654**	**336,926**	**400,835**
TENNESSEE						
Slaves/Blacks	77,485	93,830	120,682	169,474	225,804	221,452
Free/Whites	132,976	155,305	189,740	217,137	299,794	344,841
Total	**210,461**	**249,135**	**310,422**	**386,611**	**525,598**	**566,293**
TOTAL AREA						
Slaves/Blacks	146,035	203,011	243,949	384,737	536,503	616,477
Free/Whites	205,035	227,123	278,725	331,439	438,127	510,696
Total	**352,174**	**430,134**	**522,674**	**716,176**	**974,630**	**1,127,173**

[Special Note: The column for "Slaves/Blacks" only includes slaves in 1850 and 1860; it includes all blacks from 1870 to 1920; the column for Free/Whites, includes whites and free blacks in 1850 and 1860; and whites only from 1870 to 1920. Source: Bureau of the Census, *Eighth Census, Population of the United States in 1860* (Washington: Government Printing Office, 1864), pp. 18, 19, 270–271, and 466–167; Ninth Census, *Population of the United States in 1870*, Vol. 1, *(GPO, 1872), pp. 13, 14, 41–43, 61–63; Tenth Census of the United States, Population June 1, 1880*, Part I., (GPO, 1883), pp. 336–337, 358–359, 370–371; *Twelfth Census of the U.S. Population*, Part 1 (GPO, 1901), pp. 530, 545, 556–557; *Fourteenth Census of the U.S.*, Population, Vol. III, 1920 (Washington: GPO, 1922), pp. 91–97, 533–540, 961–969; *Fourteenth Census of the U.S. Population*, Vol. III, 1920, (GPO, 1922), pp. 91–97, 533–540, 961–969.]

TABLE III
POUNDS OF COTTON, 1850 TO 1920

	1850	1860	1870	1880	1900	1920
Arkansas	3,263,600	17,382,000	17,399,000	44,467,000	62,722,000	109,618,000
Mississippi	51,321,600	138,195,000	65,144,000	141,924,000	210,215,500	245,487,500
Tennessee	45,044,800	77,562,500	58,049,000	123,269,000	93,142,000	105,698,500
Total	99,630,000	231,139,500	140,592,000	306,660,000	366,079,500	460,804,000

TABLE IV
BALES OF COTTON, 1850 TO 1920

	1850	1860	1870	1880	1900	1920
Arkansas	8,159	34,764	34,798	88,934	125,444	219,236
Mississippi	128,304	276,390	130,288	283,848	420,431	490,975
Tennessee	112,612	155,125	116,098	246,538	186,284	211,397
Total	249,075	466,279	281,184	619,320	732,159	921,608

Bureau of the Census, *Eighth Census of Agriculture of the United States*, 1860 (Government Printing Office: Washington, 1864), pp. 8–9, 84–85, and 132–133; *Ninth Census of Agriculture*, 1870 (GPO, 1872), p. 102, 186, 245, 247; *Tenth Census of Agriculture*, 1880 (GPO, 1883), pp. 214, 231, 241; *Twelfth Census of Agriculture*, 1900, Crops and Irrigation, Part II (Washington: GPO, 1902), pp. 430, 432, 434; *Fourteenth Census of Agriculture*, Vol. 5, Pt. 2, Southern States, 1920 (Washington: GPO, 1922), pp. 464–472, 538–545, 575–581.

Ephemeral Plantations: The Rise and Fall of Liberian Coffee, 1870–1900

Stuart McCook

In the long nineteenth century, global demand for tropical commodities expanded at an unprecedented pace, mostly to fuel the demand from industrializing Europe and North America. To meet this demand, large new areas of the global tropics were opened to the production of commodities. The thirst for tropical commodities fueled European imperial expansion in Africa, Asia, and the Pacific. Likewise, tropical commodity exports fueled the economic growth of the newly-independent republics of Latin America. To some extent, the nineteenth-century commodity booms were simply an extension of the plantation complex that shaped tropical commodity production since the sixteenth century. But by the nineteenth century, some of the key economic and ecological underpinnings of the plantation complex had begun to change. Planters and plantations had to adapt to ensure that their crops continued to grow, and that they continued to pay.

Economic historians argue that the fate of these tropical economies was shaped by the "commodity lottery," a term which suggests that the exportable commodities produced in a given area were largely determined by "geography and chance." According to this model, "differences in later economic development were a consequence of the economic, political and institutional attributes of each commodity"[1] It is as if the landscape were a lock, and the commodity a key that would unlock the landscape's economic potential. This view treats both the commodities and their landscapes as essentially fixed and unchanging. A closer look at commodity production and trade—especially from an environmental perspective—suggests that the relationship between agricultural commodities and their landscapes was more volatile and also more deliberate than that. Constructing a viable commodity chain involved navigating complex and dynamic relationships between people, or-

1 Christopher Blattman, Jason Hwang and Jeffrey G. Williamson, "Winners and Losers in the Commodity Lottery: The Impact of Terms of Trade Growth and Volatility in the Periphery 1870–1939," *Journal of Development Economics* 82, no. 1 (January 2007): 160.

ganisms, and their enviornments, and also between local communities and global economies. The ecological relations of the plantation complex in the nineteenth century were quite volatile. At some levels, new commodity frontiers did follow the earlier models. Planters (or rather their labourers) opened new commodity frontiers by introducing exotic plants, and clearing tropical forests to establish (comparatively) homogeneous landscapes devoted to commodity production.

Likewise, "chance" had only a small role to play in determining which commodities grew where. Constant innovation was necessary to increase—or even to sustain—commodity production. Planters constantly searched for new commodities—or new varieties of existing commodities—that were adapted to the changing economic and ecological circumstances. In the nineteenth century, they initiated an unprecedented wave of global bioprospecting. There were a few signal successes, where plants fulfilled the requisite economic and ecological criteria. The transfer of rubber from Brazil to Malaya, of tea from China to India, and cacao from the Americas to West Africa are three of the most significant transfers in this period. These transfers were not the result of 'chance,' (or of 'geography' in any simple way) but rather the deliberate and systematic efforts of people and institutions. By the mid-nineteenth century, it was comparatively easy to move seeds and seedlings around the globe—much easier than it had been a century before. But the main challenge was to ensure that the introduced plants could be turned into viable commodities in their new ecological and economic contexts; very few were.[2]

The story of commodification is usually presented as a one-way story, in which something that previously had no value acquires it. In fact, the process of commodification can be transient—some commodities have value for brief moments of time, and then lose it. For example, the history of caffeinated beverages is usually (and understandably) presented as the history of coffee (always implicitly Arabica coffee), cacao, and tea. But these three crops are only a small selection of the plants that can be used to produce caffein-

2 Ross W. Jamieson, "The Essence of Commodification: Caffeine Dependencies in the Early Modern World," *Journal of Social History* 35, no. 2 (Winter 2001): 269–294; Londa Schiebinger, *Colonial Botany: Science, Commerce, and Politics in the Early Modern World* (Philadelphia: Univ. of Pennsylvania Press, 2005); L. L. Schiebinger, *Plants and Empire: Colonial Bioprospecting in the Atlantic World* (Cambridge: Harvard Univ. Press, 2004); Stuart McCook, "The Neo-Columbian Exchange: The Second Conquest of the Greater Caribbean, 1720–1930," *Latin American Research Review* 46, no. S (2011): 11–31.

ated beverages. Other caffeinated drinks such as *yerba maté* (from the plant *Ilex paraguairiensis*), and *guayusa* were consumed by indigenous groups (and later by Europeans) in South America, but never transplanted, produced, or consumed on a large scale outside of their native area. On the production side, argues Ross Jamieson, these drinks did not become global because they were not well-suited to plantation style production. On the demand side, other drinks (especially coffee and tea) already occupied the same social niche.[3]

The long nineteenth century saw a renewed and intensified quest for new tropical commodities, and new varieties of existing commodities. There were some spectacular successes, and also some significant failures. One of these failures was Liberian coffee, which was poised to become a global commodity in the 1880s and 1890s, but which ultimately failed. The introduction of new tropical commodities, both to producers and consumers, was a complex and difficult task akin to picking a combination lock. The success of a commodity is determined by a set of contingent and changing problems and opportunities, each of which is like a tumbler in a lock. To open the lock, it is necessary to successfully pick a whole set of tumblers; if even one of the tumblers is not picked, the lock does not open. For any given commodity, the tumblers in this lock involve complex combinations of environmental, economic, scientific, social, cultural and other factors. In the late nineteenth century, it looked as if some planters around the world had opened the "lock" on a new tropical commodity: Liberian coffee (known botanically as *Coffea liberica*). The case of Liberian coffee illustrates the opportunities and challenges of global plant transfers in this period. For a few brief decades in the late nineteenth century, it seemed that Liberian coffee was poised to become a new global plantation commodity. Some argued that Liberian coffee would even replace Arabica coffee in large parts of the tropics, and also allow for the opening of new coffee pioneer fronts. Liberian coffee had the backing of powerful individuals and institutions around the world. And yet, within the space of a decade, this trajectory was cut short by changing economic and ecological circumstances. But ultimately, like unsuccessful safe-crackers, they failed to hold some of the tumblers open, and Liberian coffee failed as a global commodity.

3 Jamieson, "The Essence of Commodification," 279.

The Origins of a Global Commodity: Liberian Coffee From Forest to Farm

Before the mid-nineteenth century, the global coffee industry had been based on a single species, *Coffea arabica,* or Arabica coffee. Arabica coffee is a 'mountain coffee', in the words of French geographer François Bart, native to the more temperate highlands of southwestern Ethiopia and northern Kenya, at altitudes of 450m (1500 feet) and higher. Outside its native range, Arabica coffee grew best in similar highland landscapes, such as those of the Dutch East Indies, and the island Caribbean, which produced much of the world's Arabica before the early nineteenth century. In the nineteenth century, new Arabica pioneer fronts opened up in the Indian Ocean Basin, especially in Ceylon, the Dutch East Indies, and India. They also opened up on the mainland Americas, in Central America, Colombia, Venezuela, and—above all—Brazil. In many places, Arabica pioneer fronts gradually expanded to lower altitudes, to take advantage of available land. In doing so, they pushed Arabica coffee's ecological limits, taking it into areas that were hotter and more humid than ideal for the plant.

There were other species of *Coffea* much better adapted to these hot and humid lowland environments. Many of these other species were native to the lowland forests of central and Western Africa. Until the late eighteenth century, none of these other species was known to Western scientists or traders, and none of them was traded globally. Europeans first encountered these new species of *Coffea* as they set up permanent trading establishments on the coast of West Africa.[4] One of these species, later known as Liberian coffee, grew wild in a comparatively small native range on the West African coast between what are now Sierra Leone and Liberia, and stretching to the Congo River Basin in the interior.[5] Historically Europeans had discovered new tropical commodities through local intermediaries, but in this case there

4 Jamsheed Ahmad and S. Vishveshwara, "*Coffea Liberica* Bull Ex Hiern: A Review," *Indian Coffee* (1980): 29–36; Jean Nicolas Wintgens, *Coffee: Growing, Processing, Sustainable Production: A Guidebook for Growers, Processors, Traders and Researchers* (Weinheim et al.: Wiley-VCH, 2004); Aaron P. Davis et al., "An Annotated Taxonomic Conspectus of the Genus *Coffea* (rubiaceae)," *Botanical Journal of the Linnean Society* 152, no. 4 (2006): 465–512; F. Bart, "Café Des Montagnes, Café Des Plaines," *Etudes Rurales* no. 2 (2008): 35–48.

5 J. Berthaud, "L'origine et La Distribution Des Caféiers Dans Le Monde," in *Le Commerce Du Café Avant L'ère Des Plantations Coloniales. Espaces, Réseaux, Sociétés (XVe–XIX^esiècle)*, ed. Michel Tuchscherer (Cairo: Institut français d'archéologie orientale, 2001), 361–370.

is no evidence that the Africans who lived in West Africa ever cultivated or consumed any part of this plant—it was simply one other forest plant among thousands. Virtually none of the lowland species of coffee had vernacular names in the local African languages, suggesting that they were not important in the local diets or economies.[6] The first European to document the plant now known as Liberian coffee was the Swedish botanist Adam Afzelius—a disciple of Linnaeus's and an employee of the Sierra Leone Company. Afzelius discovered the plant growing in Sierra Leone in 1792. He never published his botanical description of the plant, however, and the coffee apparently remained unknown to Western botany, and to the global coffee trade.[7]

By the early nineteenth century, some people had begun to see economic potential in Liberian coffee. It had many agronomic advantages over Arabica coffee. Liberian coffee plants were much larger than Arabica coffee—unpruned Liberian trees could grow thirty or forty feet high. Liberian coffee is "more prolific [than Arabica]," wrote the coffee trader Francis Thurber in 1881, "its berries, of varying size, being frequently much larger than those obtained from C. Arabica." The berries of Arabica trees were seldom longer than half an inch, while those of Liberian coffee were almost twice as long. At first glance, then, the Liberian coffee tree promised to be much more productive than the Arabica tree. Also, the fruit of the Liberian tree "does not drop from the tree when ripe, as in the case of Coffea Arabica." This botanical feature made Liberian coffee attractive from the farmer's perspective, since there was less pressure to harvest the trees in a short window of time between when the fruit ripened and when it fell to the ground, as with Arabica.[8]

At the same time, the fruit of the Liberian trees also presented several significant challenges. The coffee beans were difficult to process using existing techniques, since their covering was "hard, fibrous, and rather rough," making it difficult to separate the skin and pulp of the fruit from the beans.[9] Second, the Liberian plant was also much more variable than Arabica coffee, which presented a challenge in terms of producing a coffee whose taste was as consistent as Arabica. But this variability also presented an opportunity. "If there are several well-marked varieties," observed the British botanist Daniel

6 Auguste Chevalier, *Les Caféiers Du Globe* (Paris: Paul Lechevalier, 1929), 7.

7 Chevalier, *Les Caféiers Du Globe*, 75.

8 Francis Thurber, *Coffee: From Plantation to Cup. A Brief History of Coffee Production and Consumption* (14th ed., New York: American Grocer Publ. Association, 1887), 107–108.

9 Thurber, *Coffee: From Plantation to Cup*, 107–108.

Morris, "it would be well to propagate and distribute those only which exhibit a vigorous and healthy growth of wood and produce the largest crops."[10] Finally, and perhaps most importantly, Liberian coffee had a different taste than did Arabica, and it was not clear that consumers accustomed to drinking Arabica would accept this new taste.

The plant's agronomic and economic qualities were well-matched to the needs of the fledgling Liberian state. In the 1820s, former slaves from the United States established colonies in Liberia, under the auspices of the American Colonization Society. From its inception Liberia was a stratified society, with the African-American colonists forming an elite and dominating the African locals. Most of the colonists—former slaves from the United States—had little experience with tropical agriculture. These distinctive political origins shaped the course of its economic development over the nineteenth century. From its beginnings as an independent state, Liberia was poor. Capital was scarce, labour was expensive, and it did not have any advantages of landscape or climate that would attract significant foreign investment. Furthermore, its population was neither dense enough nor pliable enough to attract foreign interest. Disease was also rampant; according to one estimate, almost half of new settlers died within the first year after they arrived. Cumulatively, these factors made it difficult to establish plantations of any kind.[11]

The early settlers of Liberia must have noticed these trees growing wild in the forests around Monrovia, and begun harvesting coffee—perhaps at first for their own consumption. One of the earliest published discussions of coffee production in Liberia comes from the pen of Jehudi Ashmun, the US representative to Liberia. In 1827, Ashmun wrote that there was so much wild coffee growing along the Liberian coast that production "was so great as to render it in time, even if no new plantations are made, an important article among the exports of the Colony." A domestic market for coffee grew in Liberia, with the internal price being roughly 5 cents a pound. Nonetheless, "the time of colonists is too valuable" wrote one observer in 1827, "to be spent in picking coffee." The expansion of coffee frontiers in Liberia took place slowly. The new colonists were more interested in earning a living through trade, rather than through farming. They sold US manufactures

10 Thurber, *Coffee: From Plantation to Cup*, 107–108.

11 William E. Allen, "Historical Methodology and Writing the Liberian Past: The Case of Agriculture in the Nineteenth Century," *History in Africa* 32 (2005): 21–39.

and goods in the interior of West Africa, and in turn brought products of the region for sale to the US.

Ships returning to the United States carried sacks of Liberian coffee with them. The "quantity cured every year increases," wrote Ashmun, "and several trading vessels have taken part of what should have come to our settlements."[12] Nonetheless, Liberian coffee gained some market in the US. In 1828, Lott Cary—a former slave from Virginia who was then vice-president of the American Colonization Society—offered some 6,000 pounds of Liberian coffee for sale in Richmond, Virginia. Liberian coffee was perhaps the first 'ethical' coffee, foreshadowing the fair trade movement that was to emerge a century and a half later. Not all of this 'Liberian' coffee was cultivated in Liberia. "Coffee… is sometimes exported from Liberia;" wrote the sea captain (and astute observer) Horatio Bridge, "and doubtless the friends of Colonization drink it with great gusto, as an earnest of the progress of their philanthropic work. The cup, however, will be less grateful to their taste, when they learn that nearly all this coffee is procured at the islands of St. Thomas and St. Prince's… and entered as the produces of Liberia, *ad captandum* [to make it look good]."[13] This is, perhaps, an early example of 'greenwashing' commodities.

In the 1840s, the Liberian state started promoting export agriculture as a way of earning hard currency. The government offered incentives to convince the settlers to open up agricultural lands in the interior, particularly lands devoted to the cultivation of coffee and sugar. It offered extra land to families that would settle "more than three miles beyond town," and they offered premiums for farmers who planted coffee. The government offered a prize of $100 for the person who planted more than 5,000 coffee trees in a single year—a prize that went unclaimed, although several farmers won secondary prizes for planting more than 3,000 trees.[14] To develop tropical agriculture, the government also organized two demonstration farms, and encouraged the formation of agricultural associations. The first cultivated Liberian coffee was exported from Liberia in 1848—a total of thirteen one-

12 Jehudi Ashmun, "Coffee," *African Repository and Colonial Journal* 3 (September 1827): 218–219.

13 Horatio Bridge, *Journal of an African Cruiser: Comprising Sketches of the Canaries, the Cape De Verds, Liberia, Madeira, Sierra Leone, and Other Places of Interest on the West Coast of Africa* (New York & London: Wiley and Putnam, 1845), 44.

14 William Ezra Allen, "Sugar and Coffee; a History of Settler Agriculture in Nineteenth-Century Liberia" (Ph.D. thesis, Florida International University, 2002), 68.

hundred pound bags from the farm of one single farmer. Still, given the widespread presence of coffee trees already growing there, it is perhaps not surprising that the government officials found it difficult to induce the settlers to cultivate it.

Coffee production in Liberia remained limited and sporadic until the early 1860s, when Edward Morris, a Quaker businessman from Philadelphia, began aggressively promoting Liberian coffee both in the United States, and in Liberia itself. In the words of historian William Allen, Morris "almost singlehandedly, took coffee from the Liberian forest and placed it in the international spotlight".[15] Morris developed a mechanical coffee huller that efficiently and cleanly took the dried parchment from the coffee grain. The huller was a significant improvement over the mortars and pestles that Liberians had previously used to process their coffee. The development of this machine overcame a significant obstacle that had historically discouraged more Liberians from cultivating coffee. According to Morris, the state of the global coffee market favored Liberian coffee production. The cessation of the slave trade to Brazil after 1850, "makes it evident that the export from that country must gradually diminish," while the rising demand for coffee in the United States in Europe would ensure high prices for coffee producers in other parts of the world.[16] In retrospect, this judgment was spectacularly incorrect, but at the time it would have seemed to be sensible.

Liberian coffee remained little more than a curiosity on global markets during the 1850s and 1860s. The species accounted for a minuscule fraction of global coffee production, and it was not well known to producers outside of West Africa, or to consumers anywhere. As late as 1870, there was no particular reason to think that this situation would change. Coffee producers and consumers around the world remained content with Arabica coffee. Demand for Arabica coffee had consistently exceeded supply, and even producers of low-quality Arabicas could find a market for their crops. In the 1870s and 1880s, however, this situation began to change, opening up a niche for Liberian coffee.

15 Allen, "Sugar and Coffee," 134.

16 Anonymous, "Edward S. Morris and the Culture of Coffee in Liberia," *African Repository and Colonial Journal* 39 (December 1863): 358.

The Global 'Discovery' of Liberian Coffee: Farmers, Scientists, and Boosters, 1870–1880

In a few short years, Liberian coffee went from being a regional curiosity to a global crop. It attracted the attention of many people outside of Liberia, who hoped that the crop might be made to grow and made to pay in other parts of the world. These people—botanists, planters, and government officials, among others—hoped that Liberian coffee might help open up new coffee landscapes, and also help preserve existing ones. The global circulation of Liberian coffee was part of a larger 'neo-Columbian' exchange in the long nineteenth century. New global botanical institutions, such as the Royal Botanic Gardens in Kew, managed the long-distance transfer of plants, such as cinchona, rubber, and tea. New technological developments, particularly the advent of steamships and railroads, accelerated the pace of botanical transfers.

Liberian coffee enjoyed its greatest period of global prominence from the mid-1870s to the mid-1890s. These were generally good decades for the global coffee industry. Both demand and supply of all kinds of coffee were increasing rapidly, and for much of the century demand consistently exceeded supply. Coffee growers could usually find a buyer for almost any kind of coffee they produced. Prices in these years were so high, in fact, that coffee was regularly adulterated with other additives, from almonds to wood. From the late 1860s onward, coffee cultivation expanded at an unprecedented pace across the global tropics, with Brazil gradually assuming its position as a globally dominant producer. Demand for coffee was particularly strong in the United States, which was (and remains) the world's largest consuming market. On average, Americans consumed six times as much coffee per capita as did Europeans. In 1876, the United States consumed a third of all global coffee exports, and this share would increase in the ensuing years. Even much of the coffee produced by European colonies in this period was ultimately re-exported to the United States.[17]

New institutional and technological developments accelerated global botanical exchanges. Global botanical institutions such as the Royal Botanic Garden at Kew became significant centers for the collection and diffusion of plants across the global tropics, focusing on crops that could promote colo-

17 Mark Pendergrast, *Uncommon Grounds: The History of Coffee and How It Transformed Our World* (1st ed., New York NY: Basic Books, 1999), 61.

nial economic development. In the early nineteenth century, Kew's influence was arguably at its height, under the directorship of Sir Joseph Hooker. The Royal Botanic Gardens were involved in global transfers of other key global crops in the 1860s and 1870s—the best-known of these being cinchona, rubber, and tea. These crops were, of course, more than mere botanical curiosities. They were the engines of colonial economic development. New technological developments also helped accelerate the global diffusion of tropical plants, the development of steam shipping, the opening of canals and railroads. One innocuous but important technological invention was the Wardian case, a portable greenhouse that made the global shipping of live plants much easier than it had been previously. Other similar innovations soon followed. The English nurseryman William Bull developed a patented case, measuring 20 x 30 inches, for example, which could contain about 250 seedlings of Liberian coffee. Although the case thus packed was quite cramped, the seedlings "travel perfectly well, and have reached Queensland in good condition."[18]

All of these processes facilitated the circulation of Liberian coffee beyond West Africa. Europeans in West Africa sent live seeds and seedlings of Liberian coffee to Europe, in Wardian cases or other containers. The first shipment of Liberian coffee was sent to Kew in 1872, by John Pope Hennessy, the Irish governor of British West Africa. The nine live plants, however, perished in transit. That same year, Kew also received a shipment of Liberian coffee seeds from Accra, which were successfully cultivated in Kew's tropical greenhouses.[19] Other national botanical gardens, such as the Hortus in Leiden and the Muséum in Paris, also received Liberian coffee plants and seeds from West Africa.[20] Seedlings of Liberian coffee could be found in the greenhouses of the Museum of Paris as early as 1870.[21] Liberian coffee also reached Europe

18 G. A. Crüwell, *Liberian Coffee in Ceylon; the History of the Introduction and Progress of the Cultivation up to April, 1878* (Colombo: A.M. & J. Ferguson, 1878), 119, 154; Lucile Brockway, *Science and Colonial Expansion: The Role of the British Royal Botanic Gardens* (New Haven: Yale University Press, 2002); Richard Drayton, *Nature's Government: Science, Imperial Britain, and the "Improvement" of the World* (New Haven: Yale University Press, 2000).

19 Anonymous, "Liberian Coffee," *Bulletin of Miscellaneous Information (Royal Gardens, Kew)*, no. 47 (November 1890): 245–253.

20 P. J. S. Cramer, *A Review of Literature of Coffee Research in Indonesia*, Miscellaneous Publications Series, No. 15 (Turrialba: SIC Editorial, Inter-American Institute of Agricultural Sciences, 1957); Chevalier, *Les Caféiers Du Globe*, 77.

21 Chevalier, *Les Caféiers Du Globe*, 77.

through private botanical networks. In 1872, the same year that Liberian coffee was shipped to Kew, William Bull's tropical nursery in Chelsea (suburban London) also received a shipment of Liberian coffee from West Africa. It was William Bull who gave the plant its accepted botanical name, *Coffea liberica*, in his 1874 catalogue *A Retail List of New, Beautiful, and Rare Plants.*[22]

The Royal Botanic Gardens at Kew, Bull's nursery in Chelsea, and Irvine's nursery in Brazil quickly began producing and distributing Liberia seedlings on a large scale. In just three years, between 1872 and 1875, Liberian seeds or seedlings were sent from London to the Bahamas, Bermuda, Dominica, Jamaica, New Granada (Colombia), Montserrat, and Rio de Janeiro (Brazil), and in the eastern hemisphere to Mauritius, Natal, Ceylon, Bangalore, Calcutta, Madras, and Java.[23] Further global introductions continued well into the 1890s. Liberian coffee was often imported into the same region several times, by different people. Sometimes, the same planter would request plants both from Kew and from William Bull, on the assumption that each organization would have a slightly different variety of the plant. In the receiving areas, the Liberian coffee was planted in acclimatization gardens or experimental plots, to see how well it adapted in its many new environments. "Liberian coffee, its gigantic size, its peculiarities as to climate, and its great promise," wrote Sir Joseph Hooker, the director of Kew, "were household words from Brazil to Burmah, from Jamaica to Ceylon; from Calcutta to Queensland."[24] In the 1880s, Bull supplied Liberian coffee to planters in India, when they felt that they needed 'fresh blood' in their coffee farms.[25] In addition to institutional exchanges through botanical gardens and nurseries, individual planters also circulated Liberian coffee over long distances. For example, Liberian coffee was first introduced to Malaya either by a French or a British coffee planter, although there is some dispute as to who has the first claim.[26]

---

22 William Bull, *A Retail List (catalogue) of New, Beautiful and Rare Plants (and Orchids)* (London, 1874), 94.

23 John Reader Jackson, *Commercial Botany of the Nineteenth Century* (London: Cassell and Company, 1890).

24 D. Morris, *Notes on Liberian Coffee, Its History and Cultivation* (Kingston, Jamaica: Government Printing Establishment, 1881), 2.

25 Planter [pseudonym], "Indian Experiences: The Coffee Leaf Disease," *The Tropical Agriculturist* 7 (September 1, 1887): 199.

26 James C. Jackson, *Planters and Speculators: Chinese and European Agricultural Enterprise in Malaya, 1786–1821* (Kuala Lumpur and Singapore: University of Malaya Press, 1968), 185.

The sale and distribution of seeds and seedlings of Liberian coffee was good business. Demand from planters was high. In 1877, William Bull in London was selling seedlings of Liberian coffee at the (high) price of 50 guineas per hundred plants. He had no seed to sell at that point, presumably because it was more profitable to plant all the seeds and cultivate the trees, which would then produce more seed. In Java, in 1877, the first seeds of Liberian coffee produced in the Economic Garden there sold for almost their weight in gold. The Dutch botanist P.J.S. Cramer writes that the first batches of Liberica seed available in Java fetched very high prices—as much as 2.5 florins per seed. In Ceylon, seedlings of Liberian coffee sold for 500 rupees for 1,000 plants. Such high prices would, presumably, have limited Liberian coffee to comparatively affluent coffee producers.[27] In Jamaica, the Castleton Gardens were unable to meet the local demand for seeds and seedlings. At first, coffee planters in Liberia sought to participate in the global sale and distribution of Liberian coffee seed and seedlings. Some coffee planters went straight to Liberia to get their coffee, bypassing the European intermediaries. The Rotterdam merchant Henrick Muller, who owned large plantations in Grenville, Liberia, likely sent *C. liberica* plants directly from Liberia to Java.[28] In a better-documented case, several coffee planters from Ceylon sailed directly to Liberia to learn about how best to cultivate *C. liberica* and also to collect seeds and seedlings. By 1877, planters in Liberia reported that they could not keep up with the demand for seeds and seedlings from Ceylon, Java, and Natal.

Before the 1870s, Liberia had a default monopoly on Liberian coffee; few other people were interested in it. But as global interest in Liberian coffee grew during the 1870s, Liberia lost its effective monopoly on the germplasm, and also on the cultivation of Liberian coffee. By 1878, the Liberian government seems to have become concerned with the excessive exportation of coffee seeds and seedlings. From that year forward, they instituted a ban "prohibiting the export of [Liberian coffee] seed after the 1st of March." Perhaps this was a response to the intention of British planters in Ceylon to introduce some 18 million Liberian seeds into Ceylon. The ban reflected the Liberian government's alarm at this project, "seeing only now that we have taken the

27 Cramer, *Coffee Research in Indonesia*; Crüwell, *Liberian Coffee in Ceylon*, 142–146.

28 V. Boutilly, *Le Caféier de Libéria; Sa Culture et Sa Manipulation*, Bibliothèque de La Revue Des Cultures Coloniales (Paris: Augstin Challamel, 1900); Cramer, *Coffee Research in Indonesia*.

bread out of its mouth."[29] The Liverpool trader James Irvine—who exported coffee plants from Liberia to other parts of the world—criticized the "stupid Legislature" of Liberia for this ban, although he also noted that, as far as Ceylon is concerned, this ban "does not seem to be a matter of much moment." In any case, as an article in the *Ceylon Observer* noted that "the limited spot of West African territory called Liberia possesses no natural monopoly of what, for convenience's sake, has been called Liberian coffee."[30] Certainly, if their goal had been to preserve a monopoly on the Liberian coffee plant, and prevent its global circulation, then they were far too late. If their goal was simply to preserve domestic stocks of Liberian coffee plants, then the ban was much more understandable.[31] Still, Liberia continued to be the principal producer of Liberian coffee in Africa. In 1878, a group of Liberian planters exported 90,000 pounds of coffee to New York, at a price of 24 cents a pound. By the 1880s, some of the leading planters in Liberia produced as much as 18,000 pounds of coffee in a good year, and many smaller planters (mostly settlers)—especially around the Atherton district—were earning a good living through coffee production. Some of the local Africans also began to take up coffee cultivation, which they learned about through intermarriage with the settlers, or in the mission schools [32]

Europeans circulated Liberian coffee across Africa in the 1870s and 1880s, although attempts to establish plantations outside of Liberia had little success. Christian missionaries introduced Liberian coffee to new parts of Africa, just as they had done with Arabica coffee. In the early 1870s, French missionaries introduced Liberian coffee to the Tabou region of western Ivory Coast adjacent to Liberia. Somewhat later, Scottish missionaries introduced the crop to British Central Africa (now Malawi). These missions usually aimed to encourage smallholder coffee cultivation among Africans. But some Europeans also attempted to cultivate it as a plantation crop. The French colonial official Arthur Verdier established the first plantation of Liberian coffee in Cote d'Ivoire in the early 1880s, using seeds brought from Monrovia.[33] These efforts remained relatively small, and outside of Liberia, however, the

29 Crüwell, *Liberian Coffee in Ceylon*, 112.

30 Quoted in Crüwell, *Liberian Coffee in Ceylon*, xxxiv.

31 Crüwell, *Liberian Coffee in Ceylon*, xx.

32 Allen, "Sugar and Coffee," 143–164.

33 Louis Cordier, "Les Objectifs de La Sélection Caféière En Côte d'Ivoire," *Café, Cacao, Thé* 5, no. 3 (1961): 148; Anonymous, "Botanical Enterprise in British Central Africa," *Bulletin of Miscellaneous Information (Royal Gardens, Kew)*, no. 104 (1895): 148.

production of any kind of coffee for export in Africa remained quite limited until the early twentieth century.

Liberian coffee was also distributed across the Americas, but likewise enjoyed little commercial success outside a few small enclaves. British botanists and colonial officials promoted Liberian coffee as a peasant crop for the lowlands of the West Indies. But, beyond British and Dutch Guyana, few cultivators showed any interest. Sugar dominated agricultural production in the lowlands, and producers in the highlands preferred to grow Arabica coffee. On the mainland of Central and South America, Liberian coffee was introduced as a curiosity. But there was no compelling ecological or economic reason for coffee planters there to adopt Liberian coffee. The highland landscapes of Central and South America were, on the whole, well-suited to Arabica production, and the threat of coffee rust seemed quite distant. Liberian coffee plants were available for purchase in Costa Rica in the 1880s, but found little interest from planters because of concerns about its quality. Costa Rican coffee growers, like those in most of the rest of Central America, were primarily focused on producing high-quality Arabicas.[34] In Brazil, where coffee production focused on volume rather than quality, there was (apparently) no shortage of land for Arabica production. Brazil's coffee industry had also made the transition from slave labour to free labour without suffering any decline in exports—in fact the switch to free labour (mostly immigrants from Europe) helped make the coffee boom there possible. Other cash crops already occupied many of the lowland areas in the Americas where Liberian coffee might have flourished.

Liberian Coffee in the Indian Ocean Basin

Liberian coffee enjoyed its greatest success outside of Africa in pockets along the rim of the Indian Ocean Basin, in an arc of colonies stretching from Madagascar in the west to the Philippines in the east. In between these two termini, Liberian coffee was cultivated extensively in Réunion, India, Malaya, and—above all—Ceylon and the Dutch East Indies.

34 Mario Samper K., "The Historical Construction of Quality and Competitiveness: A Preliminary Discussion of Coffee Commodity Chains," in *The Global Coffee Economy in Africa, Asia and Latin America, 1500–1989*, eds. W. G. Clarence-Smith and Steven Topik (Cambridge: Cambridge University Press, 2003), 146–47.

Liberian coffee gained attention from planters because it solved two emergent ecological problems. First, it was apparently resistant to the coffee leaf rust (*Hemileia vastatrix*), a devastating fungal disease which was then spreading through the coffee plantations of Ceylon, Southern India, and Java. The fungus, which left characteristic orange spots on the underside of the leaves of the coffee plants, stunted the growth of coffee cherries, and in severe instances could lead to the complete defoliation of the tree. The effects of this fungus were particularly devastating in the hot and humid lowlands, where in some years the disease was so severe that farmers would lose their entire crop.[35] Liberian coffee prospered in the very regions where the coffee rust appeared most severely. "As the coffee disease and the borer continue their ravages in the coffee plantations of Madras and Ceylon," reported a botanist at the Kew Gardens, "it is much hoped, though scarcely to be expected, that the Liberian species may prove less liable to the attacks of the insect or the fungus."[36] In the field, it proved to be somewhat susceptible to the rust, but nonetheless appeared to resist attacks better than did Arabica coffee. The infected trees continued to produce large quantities of cherries.

The second ecological problem that Liberian coffee addressed was the challenge of opening new coffee frontiers, into areas unsuitable for Arabica. The British botanist Daniel Morris noted that while Arabica coffee was only suitable for cultivation "in hilly and mountainous districts," the Liberian coffee grew well "in the immediate neighbourhood of the sea and at considerable distances from it. It is to this fact that the Liberian coffee owes its importance at the present time as an economic plant of great value…." Morris predicted that "should its robust habit and prolific character be maintained in other areas its systematic cultivation promises to become one of the most successful of tropical enterprises" in areas where Arabica coffee could not be cultivated.[37] Another advantage of Liberian coffee is that it did well on exhausted soils, which could no longer support the profitable cultivation of Arabica coffees.

The Liberian plant also seemed to offer a way of addressing problems with labour. Unlike Arabica coffee plants, Liberian coffee plants flowered continuously. The production of Liberian coffee, therefore, was not seasonal. The ripe cherries of Liberian coffee remained on the tree for a long period of

35 Stuart McCook, "Global Rust Belt: *Hemileia vastatrix* and the Ecological Integration of World Coffee Production Since 1850," *Journal of Global History* 1, no. 02 (2006): 177–195.
36 Anonymous, "Liberian Coffee," 243.
37 Morris, *Notes on Liberian Coffee*, 2.

time. "The great advantage of the fruit hanging until it can be picked", wrote a planter in British Guyana, "may be judged from the fact that the opposite condition in Arabian coffee had a good deal to do with the coffee industry lapsing here fifty years ago. The supply of labour after emancipation was precarious, and coffee cultivators could not obtain it when their crop was ripe at a price that would leave any margin of profit. Hence fruit fell to the ground and rotted, and the enterprise had to be abandoned."[38] Liberian coffee gave estate planters some more breathing room and more security. They did not have to hire their labour in a short period of time, but could space out the harvest over a longer period. From the producers' perspective, then, Liberian coffee solved several key problems, and also offered new opportunities. On the consuming end, however, Liberian coffee still faced some challenges.

European planters in Ceylon were the first to attempt to take up the large-scale cultivation of Liberian coffee on a large scale. The coffee rust epidemic had broken out there in 1869, and in the 1870s spread through virtually every coffee plantation of Ceylon. When Kew began distributing seeds and seedlings of Liberian coffee in 1872, then, planters in Ceylon were particularly eager to get some. They had not been so eager two decades earlier. Early in 1851, apparently, an unidentified "Quaker firm with interests in Liberia" [presumably the American Morris] had offered some Liberian coffee to his colleague in Ceylon. While the planter was impressed by the coffee's quality, he did not bring it to Ceylon at that point because the "low country [of Ceylon] was universally regarded as unhealthy, and there was an *abundance* of cheap land in the fine mountain regions and very little inducement to extend coffee cultivation even there."[39]

By the 1870s, however, the expansion of highland cultivation, and the onset of the disease, made Liberian coffee more attractive to Ceylon planters. Coffee-planters across Ceylon scrambled to obtain some Liberian coffee—which appeared to be rust-resistant. Later in the 1870s, coffee planters disputed, in the pages of the *Ceylon Observer*, who had been the first to introduce Liberian coffee to Ceylon. These debates over priority reveal that Liberian coffee was introduced to Ceylon *many* times in the early 1870s, through a variety of channels.[40] These multiple introductions made sense:

38 Anonymous, "Liberian Coffee," 252.

39 Anonymous, "The Ceylon Handbook and Directory for 1880–1881: Acreage Under Old and New Products on Plantations in Ceylon. The Liberian Coffee Enterprise in Ceylon", *The Tropical Agriculturist* 1 (July 1, 1883): 104.

40 For a summary of these debates, see Crüwell, *Liberian Coffee in Ceylon.*

coffee planters were aware that the Liberian coffee was much more variable than the Arabica, and so they imported many varieties of the plant, from different sources, in the hopes that they would find one well-adapted to the conditions in Ceylon.[41] Planters began cultivating Liberian coffee in Ceylon's lowlands and foothills, although it did not grow as well as Arabica in Ceylon's highlands. Just a few years after the Liberian coffee farms in Ceylon's lowlands were established, however, the plants succumbed to the rust epidemic. It is not clear whether a new, specialized strain of the disease had emerged, or whether Ceylon's uniquely humid conditions provided the ideal environment for the disease to propagate. But Ceylon's Liberian coffee boom ended almost as suddenly as it had begun.[42]

Liberian coffee enjoyed its greatest success in the Netherlands East Indies. The colony's botanical garden at Buitenzorg first imported Liberian coffee to Java in 1875 or 1876. It was cultivated primarily in the humid lowlands, below 700m above sea level. The Liberian coffee plants responded to the rust epidemic differently in the Dutch East Indies than they did in Ceylon. While the plants in Ceylon ultimately succumbed within a year or two, those in the Dutch East Indies continued to produce comparatively large quantities of coffee. Producers in the Dutch East Indies found a market for Liberian coffee in the United States. It was marketed in the United States as "Liberian Java," a denomination used so that consumers would associate this coffee with the high-quality Java coffees (all Arabicas) that were prized in the US market. In the 1890s, there was a small boom in the production of Liberian coffee. In 1893, government production of Liberian coffee was 607 piculs [37,000 kg]. By 1901, production reached a peak of 29,849 piculs [1.84 million kg]. In 1896, the planters association in West Java proclaimed that in their region "Liberian coffee has taken the place of Arabian coffee almost completely. There is a strong likelihood that, as a consequence, this part of our beautiful island will reach the same degree of prosperity it enjoyed during the flourishing period of Arabian coffee."[43]

41 William Jardine and Henry Trimen, "Disease-Resisting Liberian Coffee Trees," *The Tropical Agriculturist* 1 (June 1, 1882): 1073.

42 James L. A. Webb, *Tropical Pioneers: Human Agency and Ecological Change in the Highlands of Sri Lanka, 1800–1900* (Athens: Ohio University Press, 2002).

43 Quoted in Cramer, *Coffee Research in Indonesia*, 107. See also P.J.S. Cramer, "L'influence de l'Hemileia Vastatrix Sur La Culture Du Café Au Java," *L'Agronomie Tropicale (Brussels)* 2, no. 2 (November 1910): 389–93.

By the early 1890s, the Liberian coffee appeared to have a promising future as a plantation crop, particularly in the Dutch East Indies as well as a host of smaller producers which, for one reason or another, could not produce Arabica coffee. In many respects, Liberian coffee seemed to be well-suited to the prevailing economic and ecological issues facing coffee planters, particularly in Asia and the Pacific and to a lesser extent Africa. It could grow in lowland landscapes, allowing in principle for the expansion of coffee frontiers. It required less labor than Arabica coffee. It showed at least partial resistance to the leading epidemic disease of coffee, and to one of the worst insect pests of coffee. For many coffee farmers, Liberian coffee appeared to be a godsend. The main challenge facing them was to convince *consumers* to develop a taste for Liberian coffee.

The Challenge of Taste: Creating Global Demand for Liberian Coffee

While Liberian coffee offered a lot of potential advantages to coffee producers, its taste remained a major stumbling block for consumers. Coffee experts all agreed that Liberian coffee tasted different than Arabica coffee. But they could not agree on whether or not it tasted *better* than Arabica coffee. The issue of taste in coffee is complex; there was (and is) no objective measure for coffee quality (although there may be objective measures for differentiating coffee tastes).[44] "Liberian coffee," wrote a British journalist in Ceylon, "which for nearly one month the writer has drunk daily, has a peculiar flavour, but to our taste it was far from being objectionable." Liberian coffee was, according to a Dutch journalist visiting one of Java's first large Liberian estates, "more fragrant than the common variety [Arabica], but when it is tried for the first time there may be something peculiarly unpleasant about it." Still, all was not lost: "after a couple of days this difference is no longer perceptible."[45]

44 Samper K., "The Historical Construction of Quality and Competitiveness: A Preliminary Discussion of Coffee Commodity Chains."

45 Frederick Wellman, *Coffee: Botany, Cultivation and Utilization* (London: Leonard Hill, 1961), 76–79; Cramer, *Coffee Research in Indonesia*; Thurber, *Coffee: From Plantation to Cup*, 107–116; James H. Smyth, "Liberian Coffee," *The Tropical Agriculturist* 2 (1882):

There was also some question about whether or not the problems with Liberian coffee's flavor were inherent to the bean itself, or whether they were the result of poor harvesting, cultivation, and preparation. The coffee's flavor had much to do with how the beans were processed and brewed. "The coffee we drank was strong, hot and with the aroma still in it, which you do often miss in coffee when not prepared by skillful hands, from whatever coffee the country may come."[46] Not all Liberian coffee was prepared with equal attention; some London buyers characterized Liberian coffee as "undrinkable." This was attributed, by one British planter in Ceylon—anxious to distinguish cultivation in Ceylon from cultivation in Liberia—as being due to the fact that "the product hitherto supplied has been gathered from uncultivated trees [in Liberia] and very probably mixed with unripe berries."[47] The coffee's final taste was also shaped by other steps in the post-harvest process—whether the beans were pulped by being dried in the sun (the dry process), or "washed" (the wet process). Consumers usually preferred the taste of washed coffees, but this process was much more cost-intensive than the dry process.

Even the advocates of Liberian coffee recognized that consumers would likely need to be educated to appreciate the bean. In 1881, some coffee dealers in London noted that Liberian coffee had considerable potential as a crop, "and will be well to the front in competition with all other descriptions, as soon as the general taste becomes educated to a fuller bodied although less delicate, flavour than now produced from Ceylon and East Indian Plantation Coffee."[48]At least initially, in the 1870s and 1880s, the consensus seemed to be that Liberian coffee tasted better than Arabica coffee, and the prices it fetched on the market reflected that. Boosters of Liberian coffee regularly drew an analogy between Liberian coffee and tea. They noted that British consumers had become accustomed to drinking teas from two distinct origins (China and Assam) with distinct flavors, and there was room in the market for both. They hoped that the coffee market could, likewise, have room for two species of coffee with distinct flavors. At the time, the global coffee market had not yet been fully integrated, and at least in theory this remained a possibility. Over the longer term, however, neither traders nor consumers

325–28; Anonymous, "Liberian Coffee and Tea in Java," *The Tropical Agriculturist* 1 (July 1, 1881): 159.

46 Crüwell, *Liberian Coffee in Ceylon*, 62.

47 Ibid., 150.

48 Morris, *Notes on Liberian Coffee*, 13.

ever became fully accustomed to this idea, and Arabica remained the benchmark for quality coffee [49]

The plant's popularity was given a boost by its strong showing at fairs and exhibitions. Edward Morris, a prominent coffee farmer and foreigner living in Liberia, showed Liberian coffee at the US Centennial Exhibition in Philadelphia, in 1876. The coffee won the "US Centennial Diploma Awards for Superior Coffee," and the "Medals of Awards and Diplomas of Merit, for Superior Coffee and Palm Soap", which enhanced its reputation across the United States and also globally.[50] It earned accolades from leading Americans; the eminent chemist Benjamin Silliman of Yale found the quality of Liberian coffee to be "so much superior to most coffee in common use in this country that I at once ordered a sample [from Morris]." Silliman concluded that "the coffee of Liberia is equal to any coffee that I have ever seen."[51] The coffee enjoyed a generally strong reputation for the next twenty years, winning another round of accolades at the World's Columbian Exhibition in Chicago in 1893. There, it won awards for its strength and its compactness of grain.[52]

The volatile prices for Liberian coffee in the 1870s and 1880s reflected uncertainty about how to judge its quality. As the *British Trade Journal* noted in 1877, "it is always difficult to estimate the intrinsic value of a comparatively new product, and first prices are apt to be misleading." Parcels of Liberian coffee sold for around 100/ per cwt [or 120 British pence per pound] in 1877, but "as motives of curiosity must have actuated buyers, it must not be assumed that these figures represent its real value." In short, the article concluded, these were "fancy prices."[53] During the 1880s, the price of Liberian coffee in global markets reached a new high, receiving 20 cents a pound. Buyers in New York, the main coffee market in the US, paid between 18 and 22 cents a pound for large-beaned Liberian coffee in the late 1870s.[54] Even lower prices could also be attractive to planters, reflecting the almost insa-

49 Anonymous, "Liberian Coffee Cultivation," *The Tropical Agriculturist* 1 (November 1, 1881): 461; Steven Topik, "The Integration of the World Coffee Market," in *The Global Coffee Economy in Africa, Asia and Latin America, 1500–1989*, eds. W. G. Clarence-Smith and Steven Topik (Cambridge: Cambridge University Press, 2003), 21–49.

50 Allen, "Sugar and Coffee," 141.

51 Quoted: Ibid., 136.

52 Alfred B. King, "Liberian Coffee," in *Liberia*, vol. No.1–34 (1895): 40–49.

53 Crüwell, *Liberian Coffee in Ceylon*, 148.

54 Allen, "Rethinking the History of Settler Agriculture in Nineteenth-Century Liberia," 454; Crüwell, *Liberian Coffee in Ceylon*, 109.

tiable demand for coffee of all kinds in Europe and the US. "If villainously cured and smashed coffee fetches 70/ a cwt (roughly equivalent to 17 cents a pound) in the London market, what will Ceylon-grown Liberian coffee, well cured and not smashed, fetch?.... Even at 70s. a cwt, Liberian coffee may pay better than C. Arabica grown in some of the mountains of Ceylon, just as Brazil coffee yields larger profits to the grower in that country owing to the heavy bearing capacities of the trees and the richness and fertility of the soil in spite of a much lower price and in spite of the 13 per cent export duty."[55]

Given the market conditions, Liberian coffee could be made to pay either as a luxury coffee or as a commodity coffee. In the United States, Liberian coffee initially found favor among lower-class drinkers, although the details are sketchy. One of the few glimpses we get is from the observer Boutilly, a French colonial official in Réunion and a passionate advocate of Liberian coffee, who noted that it was "originally directed exclusively to Louisiana and Kentucky." By the turn of the century, demand was located primarily in the US West. Coffee consumption in the US grew after the Civil War, when coffee had been included in the rations of Union soldiers. Liberian coffee from Ceylon found a market in the United States, although one Ceylonese planter complained that they had to make the grains of dried Liberia look *worse* in order to be marketable: "Messrs Leechman have actually been compelled to prepare this coffee specially for the American market, so that it may have, instead of the greenish blue hue of well ripened and properly prepared beans, the dirty yellow colour, which the unripe and badly prepared coffee originally received from Liberia had accustomed [American] consumers to expect."[56] The planter hoped that, over time, the American market could be convinced to buy *good* Liberian coffee.

After the Civil War, it appears that American coffee drinkers no longer drank Liberian coffee on its own, but rather as a blend, "mixed with other coffee to impart 'flavour.'[57] Again, according to Boutilly, it was "primarily consumed by a working-class population of German and Irish origin, who mix it with a high proportion of alcohol."[58] The consumption of Liberian coffee in blends reflected the changing American coffee market in the 1870s and 1880s. Before then, coffee had been sold in small lots of beans. After the

55 Crüwell, *Liberian Coffee in Ceylon*, 62.

56 Anonymous, "No. 3. The Kurunegala Road–Udapolla Estate–Liberian Coffee," *The Tropical Agriculturist* 1 (April 1, 1882): 829.

57 Crüwell, *Liberian Coffee in Ceylon*, 62.

58 Boutilly, *Le Caféier de Libéria*, 26.

Civil War, coffee was increasingly mass-produced, sold roasted, ground, and packaged.[59] Nonetheless, other pieces of evidence suggest that the US market was more diversified than that. Liberian coffee from Java was sold, of all places, by Sears and Roebuck. The 1897 Sears catalogue listed Liberian Java coffee at 23 cents a pound. This, incidentally, was just a half-cent less than the prized Old Government Java (Arabica coffee), and seven cents a pound more than ordinary Rio coffee. Java coffee also sold in England, in Holland, and in France.[60] For this window of time, then, Liberian coffee addressed opportunities and solved problems both for producers and consumers.

The Death of a Commodity: The Decline of Liberian Coffee after 1895

In just a few years, the situation of Liberian coffee changed dramatically, and after the mid-1890s it never regained the promise or prominence that it had enjoyed for the previous three decades. The Liberian coffee plant faced a perfect storm of ecological, economic, and other problems that emerged during the 1890s. Perhaps the largest of these was the general global economic crisis of the 1890s, which sent commodity prices spiraling downward. The global coffee market, in particular, was transformed by the rapid expansion of coffee production in Brazil, which tripled in the last decade of the nineteenth century. By the 1890s Brazil produced 80 percent of the world's coffee; in other words, Brazil alone produced four times as much coffee as the rest of the world's coffee producers combined. Brazil enjoyed abundant land for Arabica production, and an equally abundant supply of labour. Even during the boom years of the 1870s, the spectre of overproduction had begun to loom. The large Brazilian crops of 1878, wrote the coffee expert Francis Thurber, "made it apparent that the time when production would exceed consumption had come."[61]

It proved difficult for coffee producers in Africa, Asia, and the Pacific to compete with their counterparts in the Americas. Repeated bumper crops of

59 Pendergrast, *Uncommon Grounds*, 48–51.

60 Skyhorse Publishing, *1897 Sears Roebuck & Co. Catalogue* (New York: Skyhorse Publishing Inc., 2007), 9.

61 Thurber, *Coffee: From Plantation to Cup*. Quoted in Pendergrast, *Uncommon Grounds*, 64.

Brazilian coffee pushed supplies upward and the price of coffee downwards, hitting a low of 4.25 cents a pound in 1898. As a whole, global coffee prices remained catastrophically low for the next decade.[62] Reflecting this, the price of Liberian coffee dropped from its high of 20¢/lb to a low of 6¢/lb in the late 1890s. Between 1897 and 1900 alone, the price of Liberian coffee in the United States dropped 40 percent, largely because of the flood of Brazilian coffees on the market.[63] The French consul in Java concluded that "given the current prices of Brazilian coffee, it is impossible for planters to produce [Liberian coffee] without a financial loss."[64] It proved difficult for coffee producers in Africa, Asia, and the Pacific to compete with their counterparts in the Americas. Whereas the eastern hemisphere had accounted for about a third of global coffee production in the 1830s, by 1905 that share (which includes both Arabica and Liberica coffee) had diminished to less than 5 percent. Part of this represents a relative loss, reflecting the dramatic increase in production in the Americas. But part of it also reflects an absolute decline of production in the East. Liberian coffee was caught up in this more generalized regional decline [65]

At the same time, debates over the quality of Liberian coffee persisted. The historian William Allen argues that "Liberian coffee beans were 'inferior' primarily because they were poorly prepared, not because *C. liberica* was a flawed species."[66] The taste was a product of many factors—the plant itself, the *terroir* in which it was cultivated, and the way that it was processed, shipped, and roasted could all shape the flavour of the final product. In the Dutch East Indies, planters and botanists responded by making a concerted effort to improve the selection, cultivation, and processing of Liberian coffee, to produce a more consistent and better-tasting product. But the machinery to process the beans—in particular the specialized hullers necessary to process the hard and variable Liberian coffee fruit—was so expensive as to be beyond the reach of many small cultivators. There were some mechanical hullers in the Dutch East Indies, in India, and in Liberia. But in most places small farmers processed their coffee using the 'dry' method, which

62 Pendergrast, *Uncommon Grounds*, 68, 77–79.

63 Boutilly, *Le Caféier de Libéria*, 27.

64 Boutilly, *Le Caféier de Libéria*, 38.

65 W. G. Clarence-Smith, "The Coffee Crisis in Asia, Africa, and the Pacific, 1870–1914," in *The Global Coffee Economy in Africa, Asia and Latin America, 1500–1989*, eds. W. G. Clarence-Smith and Steven Topik (Cambridge: Cambridge University Press, 2003), 100–119.

66 Allen, "Sugar and Coffee," 194.

often harmed the taste of the final product. At a time of falling global coffee prices, few coffee farmers had the capital to invest in technologies that would improve the final taste. In the end, the questionable flavour of Liberian coffee may have limited its appeal to roasters and consumers in North America and Europe.

The economic troubles facing Liberian coffee were compounded by the new ecological problems, specifically the spread of the coffee leaf rust. From the beginning, coffee planters had recognized that Liberian coffee was susceptible to the *Hemileia vastatrix* fungus. But at first, Liberian coffee had seemed *less* susceptible than Arabica coffee. In Ceylon, the early plantings of Liberian coffee had succumbed to the rust after just a few years, and by the early 1880s coffee planters there had largely abandoned the plant. But through the 1880s and early 1890s, Liberian coffee planted elsewhere in the region still seemed to enjoy a comparative immunity to the rust. By the mid-1890s, however, it seemed that the Liberian coffee plantations in Southeast Asia 'lost' much of their resistance to the rust. In retrospect, it would appear that a new strain of the coffee rust—one specialized in attacking the Liberian coffee plants—had emerged and spread through the region, causing losses on the same scale as it had on Arabica coffee. Liberian plants in India began succumbing to the rust in 1893; the following year they began to succumb in Malaya. It had been a sporadic problem in the Dutch East Indies in the mid-1890s, and began to cause serious losses after 1897.[67]

In response to this, planters abandoned Liberian coffee on a large scale; many of them switched to cultivating rubber. Rubber trees had been introduced to Southeast Asia from Brazil in the mid-nineteenth century. In Brazil, it had proved impossible to grow rubber as a plantation crop because of a fungus native to Amazonia, the South American Leaf Blight, which was particularly severe on rubber trees cultivated as monocultures. Since this disease was absent in Southeast Asia, rubber trees prospered in a plantation environment. Former Liberica farms in Malaya, Java, and Sumatra were converted to rubber plantations.[68]

67 Cramer, *Coffee Research in Indonesia*, 43–45; George Romilly to W. T. Thiselton-Dyer, November 18, 1893, Miscellaneous Report: India Economic Products, ff. 80–81, Library and Archives, Royal Botanic Gardens, Kew; Jackson, *Planters and Speculators*, 200.

68 Warren Dean, *Brazil and the Struggle for Rubber: A Study in Environmental History* (Cambridge: Cambridge University Press, 1987); Greg Grandin, *Fordlandia: The Rise and Fall of Henry Ford's Forgotten Jungle City* (New York: Metropolitan Books, 2009); Isaac Henry Burkill, *A Dictionary of the Economic Products of the Malay Peninsula* (vol. 2, Kuala Lumpur: Ministry of Agriculture and Co-operatives, 1966).

The final death knell for Liberian coffee came in the opening decades of the twentieth century, from another species of coffee. The European 'scramble for Africa' after 1885 had propelled Europeans into the interior of the continent. There, European botanists uncovered many new species of coffee, and distributed them around the world through the same networks that Liberian coffee had followed several decades earlier. Most of these new species of coffee had no commercial value. One of them, known commercially as Robusta coffee (*C. canephora* var Robusta), however, enjoyed considerable success. Robusta coffee occupied almost exactly the same ecological and economic niches as did Liberian coffee; it had most of Liberian coffee's advantages and few of its disadvantages. It grew well in the hot and humid lowlands of the tropics, and produced a large quantity of beans. Unlike Liberian coffee, Robusta coffee was highly resistant to the coffee rust (as well as to many other diseases that plagued coffee). The first Robusta plants were introduced from central Africa to Java in 1900. By 1910, the government of the Dutch East Indies had recognized the crop's potential as a substitute for Liberian coffee, and began a large planting program in the 1910s and 1920s.[69]

From the perspective of global markets, Robusta's lack of cupping character proved to be the key to its economic success. Robusta coffee could be roasted, ground, and blended with higher-quality Arabica coffees, which would provide the aroma and flavour to the mix. In short, coffee roasters adulterated one kind of coffee with another kind of coffee. Like Liberian coffee, then, Robusta coffee generally occupied the same market niche as did the lower-grade Brazilian coffees. Across the twentieth century, Robusta cultivation spread from the Dutch East Indies across the Indian Ocean Basin, the Pacific, and Africa—helped at times by price spikes in Arabica coffee, and at times by tariff protections. By the early twenty-first century, Robusta accounted for a third of global coffee production—far more than Liberia had ever achieved.

Even after the spectacular collapse of the 1890s, Liberian coffee cultivation did not disappear entirely. There are pockets of production even to the present. In 1930, for example, the French botanist August Chevalier estimated that "a few hundred tons a year are produced." In 1930, Liberian coffee remained the only kind of coffee cultivated in Sierra Leone, Liberia, Ivory Coast, and Lagos [Nigeria].[70] It was also widely produced in the islands in

69 Wellman, *Coffee*, 80–87; Gordon Wrigley, *Coffee* (New York: Wiley, 1988), 54–58.
70 Chevalier, *Les Caféiers Du Globe*, 77.

the Gulf of Guinea of West Africa, including São Tomé and Príncipe (Portuguese colonies until 1975) and Fernando Pó (also known as Bioko), a former Spanish colony belonging to Equatorial Guinea. In Southeast Asia and the Pacific, small enclaves of Liberian coffee production persisted in Malaya (later Malaysia) and the Philippines. In the Americas, British and Dutch Guyana (later Guyana and Suriname) also continued to produce and export small quantities of Liberian coffee. Small niche markets for Liberian coffee also persisted into the twentieth century. For example in the first half of the twentieth century, Norway consumed most of Suriname harvest of Liberian coffee.[71]

In the twentieth century, however, even most of these few small pockets of Liberian coffee cultivation were abandoned. Between 1930 and 1950 Robusta cultivation gradually surpassed Liberian cultivation even in West Africa. European planters in the Ivory Coast abandoned Liberian coffee around 1928, while African planters there continued to cultivate it until 1934. Both groups of planters switched to the cultivation of Robusta coffee, which had been introduced to the Ivory Coast in 1915.[72] Even this small niche, however, was gradually squeezed out. After Norway signed a bilateral treaty with Brazil around 1960, it stopped importing Liberian coffee from Suriname.[73] Today most Liberian coffee is produced by smallholders, to meet local or regional demand. Information about it, however, is quite difficult to find. These local and regional commodity chains are worth studying in more detail, since they do not represent any kind of return to traditional coffee cultivation; the crop was exotic to almost everyone who cultivated it and they had not prior local knowledge of this crop. A study of the absorption of Liberian coffee into local agricultures, local economies, and local diets would help shed light on innovation in non-Western societies.

71 J. E. Heesterman, "Quelques Observations Sur Le Goût Du Café Libéria et Du Café Soluble En Poudre a Partir de Cette Espèce," *Café, Cacao, Thé* 7, no. 4 (1963): 326–330.

72 Cordier, "La Sélection Caféière En Côte d'Ivoire," 148.

73 Heesterman, "Quelques Observations Sur Le Goût Du Café Libéria et Du Café Soluble En Poudre a Partir de Cette Espèce," 326.

Conclusions: The Rise and Fall of a Commodity Niche

The case of Liberian coffee shows how the so-called "commodity lottery" is less random than the term suggests. The repertoire of commodities that a given area produces may be volatile, but it is not accidental. The story of Liberian coffee also highlights the complex range of factors that must operate simultaneously necessary for a plant to become a global commodity. While it was theoretically possible to cultivate Liberian coffee in a wide range of humid lowlands across the global tropics, in practice it was only viable in a few places. It flourished in a brief moment when the global coffee market and coffee prices were generally expanding faster than production, when public and private institutions diffused the plant across the globe and promoted its cultivation; and when planters felt that the plant could resolve many of their most pressing problems. At the same time, Liberian coffee only became part of larger consumption patterns in Europe or North America. It enjoyed some success among coffee traders as an exotic curiosity, but it seems there was never widespread acceptance for its distinctive taste. To the extent it did succeed, it was as an adulterant for Arabica coffee, whose taste dominated in the blend. It was a tiny niche in a coffee market dominated by Arabica coffee. Even that niche proved unsustainable, both economically and environmentally. The spread of the coffee rust made it impossible to cultivate Liberian coffee profitably through much of the Indian Ocean Basin.

More importantly, the people who produced Liberian coffee were rarely, if ever, committed to producing a single commodity. Rather, they were involved in dynamic and changing 'commodity webs,' at the intersection of many different actual and potential commodity chains.[74] As local and global conditions for Liberian coffee worsened, and new opportunities to grow other crops emerged, they switched their focus. Depending on the locale and the time, Liberian coffee was replaced by rubber, cacao, coconut palms, or other cash crops. Even those farmers who were still committed to coffee switched to Robusta, which offered many of the same advantages as Liberian coffee, with few of its disadvantages. Robusta succeeded as a global commodity where Liberian coffee did not; a global market for this 'inferior' coffee is now well established. The various stakeholders along the Robusta

74 Thomas Sikor and Pham Thi Tuong Vi, "The Dynamics of Commoditization in a Vietnamese Uplands Village, 1980–2000," *Journal of Agrarian Change* 5, no. 3 (2005): 405–428.

coffee chain were able to "pick the lock" in ways that Liberian coffee producers never were.

Finally, the case of Liberian coffee reminds us that "commodification" can be a two-way process. An object can be commodified, and then de-commodified. Its case is not isolated, either. Many of the leading plantation crops of the nineteenth century have seen their global markets shrink significantly or disappear altogether over the twentieth century. Cinchona trees, whose bark was valued as an antimalarial, was circulated and cultivated globally at around the same time as Liberian coffee, and harvested in plantations across the Indian Ocean Basin. With the development of modern synthetic medicines in the twentieth century, however, the value of cinchona bark plummeted and cinchona plantations failed. Similarly, indigo was a major plantation crop, producing a highly-prized dye. When German companies developed synthetic indigo at the turn of the twentieth century, however, cultivated indigo could no longer compete. By the 1920s, the era of indigo plantations had come to an end.[75] More recently, some scientists and journalists have expressed fear for the future of the banana industry in the face of new blights that threaten the main banana cultivar, and the limited ability to produce new cultivars that are economically viable.[76] Even the most powerful commodity chains can be vulnerable, and even the most powerful plantations can be ephemeral.

75 Prakash Kumar, *Indigo Plantations and Science in Colonial India* (Cambridge and New York: Cambridge University Press, 2012).

76 Fred Pearce, "Last Days of the Banana: The World's Favourite Fruit Is About to Disappear," *New Scientist* 18 (2003).

Between “Wild Tropics” and “Civilization”: Guatemalan Coffee Plantations as Seen by German Immigrants

Christiane Berth

“I was now in the land of coffee. In the land, where coffee grew, where everybody traded with coffee, where everybody somehow lived on coffee, where the whole economic life was dominated by coffee prices. It was obvious for me to ‘go into coffee’ as well, as it was called in Guatemala.”[1]

The German immigrant Helmuth Schmolck described coffee as the all-dominant force in Guatemala. Indeed, coffee became the country’s most important export product in the second half of the 19th century. It linked isolated regions to the world economy, changed social structures as well as landholding patterns and transformed whole landscapes. The coffee elites remained the central political force until the 1930s.

German immigrants played a dominant role in the Guatemalan coffee business. At the beginning of the 20th century, they generated roughly one third of the Guatemalan green coffee. Several German coffee producers published autobiographies or travel reports describing in detail their new working environment. These texts are an important source to examine social relations on the plantations, the perceptions of the tropical environment, the organization of coffee plantations and their travel experiences. Written in a great part during the first coffee boom until the end of the 19th century, they reflect the optimism regarding the great business opportunities. At the same time, they also convey some concerns about fluctuating coffee prices and political unrest. Further, these German accounts document the experimental phase of coffee production, when plantation owners tried to find the best way of organizing a plantation, and applied different production patterns and planting methods. But there are also subjects completely ignored, such as social conflicts, business failure, and cultural misunderstandings.

Although recently the ecological consequences of contemporary coffee production have been intensely debated, there is little research on the en-

1 Helmuth Schmolck, *Welthandel selbst erlebt*, 2nd ed (Heidelberg: Vowinckel, 1951), 69.

vironmental effects of coffee production in a historical perspective.[2] In his study on banana cultures in Honduras, the historian John Soluri suggested to work with the concept of a "commodity web" and to analyze ecological transformations, diseases and people working on the plantations as central variables within it.[3] Such an integrated history linking ecological transformations, working environments, the emergence of mass markets and the activities of foreign investors has not yet been written for coffee production either.

This article is an invitation to look at the ecological history of coffee plantation through the eyes of German immigrants. Their perspective offers new insights into the period of plantation expansion in the Alta Verapaz in general and on ecological concerns in the building-up of a plantation. Their perspective also shows how the perception of nature influenced plantation owner's decision-making. Finally, it demonstrates that their "civilizing mission" embraced the social and the ecological environment. For the Indian case, David Arnold has shown how travelers started and expanded the European appropriation of local landscapes and nature in the first half of the 19th century. European scientific explorations of the tropics were linked with the expansion of British and French plantation societies and contributed to define the tropics as a space distinct to Europe including political, cultural, and natural aspects.[4] In Guatemala, German plantation owners and geographers disseminated the idea of Guatemalan tropical nature as a wild, uncivilized space. For them, founding plantations signified converting tropical nature into organized, productive landscapes. My central hypothesis is that the abundance of land and the absence of major infections allowed German plantation owners not to be worried about soil degradation and deforestation. Their awareness for environmental change was limited to certain as-

2 Stefania Gallini, *Una historia ambiental del café en Guatemala. La Costa Cuca entre 1830 y 1902* (Guatemala: Avansco, 2009); Andrés Guhl, *Café y cambio de paisaje en Colombia, 1970–2005* (Medellín: Fondo Editorial Universidad EAFIT, 2008); Stuart McCook, "La Roya del Café en Costa Rica: epidemias, innovación y medio ambiente, 1950–1995," *Revista de Historia* 59–60 (2009): 99–117; Stuart McCook, "Global rust belt: *Hemileia vastatrix* and the ecological integration of world coffee -production since 1850," *Journal of Global History* 1 (2006), 177–195.

3 John Soluri, *Banana cultures. Agriculture, consumption, and environmental change in Honduras and the United States* (Austin, TX: Univ. of Texas Press, 2005), 226.

4 David Arnold, *The Tropics and the Traveling Gaze. India, Landscape, and Science, 1800–1856* (Seattle/London: Univ. of Washington Press, 2006), 110–115.

pects as weeds and fertilizing. In general, they perceived nature as a resource for generating profits.

The article is structured in three parts: I will first give a short overview on coffee production in Guatemala, coffee processing, and the influence of German immigrants. Based on their accounts, I will analyze how they described tropical nature and related those descriptions to their perception of the tropics in general. I will argue that dominating tropical nature was presented as part of the German pioneer work in "civilizing" Guatemalan society. Second, I will outline how the authors depicted the environment of the coffee plantations. Third, I will show how the Germans described social hierarchies and extended their civilizing mission to the indigenous workers. Finally, I will analyze their descriptions of global and local influences on the plantations.

Coffee production in Guatemala in the 19th century and the role of German immigrants

The first coffee export in Guatemala, registered in 1853, was a ray of light in an otherwise somber economic panorama. Due to the invention of chemical substitutes, the traditional Guatemalan export products indigo and cochenille entered a stage of crisis in the mid-19th century. Consequently, the Guatemalan government began to look for alternatives and finally chose coffee as an option for the export economy. From the 1850s onwards the government encouraged coffee production and enacted laws benefitting finca owners.

All over Central America, coffee production had expanded since the 1830s. Following the Costa Rican example, plantation owners applied the wet method for processing, which gave coffee a special taste which found great acceptance on European markets. Rapidly, Central American coffee occupied an important segment in the sector of quality coffees. In Guatemala, the expansion of coffee production occurred very quickly: Its share of all exports rose from 1 percent in 1860 to 36 percent in 1868.[5] In 1862, already more than 5,5 million coffee trees were growing in Guatemala. As land distribution was historically unequal, coffee production mostly took place on large plantation complexes. In 1890, 53 percent of all coffee was produced

5 Regina Wagner, *Historia del café de Guatemala* (Bogotá: Villegas editores, 2001), 51. The information refers to the value of exports in Guatemalan pesos.

on plantations with more than 100,000 trees.[6] The rapid expansion of production changed social structures in Guatemalan society fundamentally. It changed the patterns of land ownership as well as local power structures and linked formerly isolated regions to the world economy.

In the 1870s, liberal governments came to power with the agenda of strongly promoting agricultural exports. The Liberal Reforms aimed at breaking the power of the church and giving finca owners access to land and workers. President Justo Rufino Barrios (1873–1885) enacted a series of laws which induced the privatization of land ownership in the country. Since 1877, it was possible to claim land as unused ("tierras baldias") which the government then had to offer for auction. In some coffee growing regions, indigenous communities lost their land and reacted violently to the introduction of coffee production.[7]

Moreover, the Guatemalan government re-introduced colonial forms of forced labor called "mandamientos". When plantation owners demanded workers from the government, indigenous communities were obliged to provide a contingent. At the same time, seasonal workers were recruited by advance payments, often leading to indebtedness and creating dependence on the plantation owners. In the Alta Verapaz a third form, called "colonato" was quite common: Plantation owners paid workers part of their wage in cash and part of their wage in rights of land use.[8]

Another important factor for the expansion of coffee production was the local infrastructure. In the beginning, coffee had to be transported by mules on small mountain paths to the port cities. As construction of railways throughout the country made transportation faster and more efficient, coffee plantations close to a railway station garnered great advantages. The Guatemalan government built several new ports since the 1870s, for example in Champerico (1877), Livingstone (1878) and Ocós (1884). In 1880, it entered

6 Robert G. Williams, *States and social evolution. Coffee and the rise of national governments in Central America* (Chapel Hill, NC: Univ. of North Carolina Press, 1994), 64; Wagner, *Historia del café de Guatemala*, 49–56.

7 Mario Samper Kutschbach, *Producción cafetalera y poder político en Centroamérica (Colección Rueda del tiempo)* (San José, Costa Rica: EDUCA, 1998), 81–90; Wagner, *Historia del café de Guatemala*, 85–98; Williams, *States and social evolution*, 58–61; David McCreery, *Rural Guatemala, 1760 –1940* (Stanford, CA: Stanford Univ. Press, 1996), 163–164, 238–258.

8 McCreery, *Rural Guatemala*, 192–193; Williams, *States and social evolution*, 112–119; Karl Theodor Sapper, *Mittel-Amerika*, Auslandswegweiser 5 (Hamburg: L. Friederichsen & Co, 1921), 102–105.

into a contract with the Kosmos shipping line to establish a regular direct connection with Northern Germany.[9] In the Alta Verapaz, a local society was founded to construct a railway linking Cobán with Livingstone. The consortium consisted of several German coffee exporters and coffee trading houses from Hamburg.

The emerging group of German immigrants in Guatemala benefitted from this liberal policy. Since the 1840s, the growing commercial exchange had attracted Germans from Hanseatic cities to Central American coffee producing regions. In Guatemala, the German presence was strongest: Germans were involved in the coffee production, coffee export, infrastructural projects, shipping lines and banking. In any case, they could rely on cheap credits from their hometowns and the protection by diplomats from the German empire. By promoting European immigration, the Guatemalan government sought to import new technologies and infrastructure.[10] But racial considerations played an important role as well: Generally, Guatemalan society in the 19th century was deeply divided by ethnic criteria. The mestizo or white elites considered the indigenous population as an inferior group that had to be civilized by hard work.[11] Moreover, they tried to "improve" the composition of the Guatemalan population by attracting white immigrants from Europe.[12]

9 Wagner, *Historia del café de Guatemala*, 85–98; Thomas Schoonover, *Germany in Central America. Competitive Imperialism, 1821–1929* (Tuscaloosa: Univ. of Alabama Press, 1998).

10 See on the history of German immigration in Guatemala Regina Wagner, *Los alemanes en Guatemala 1828–1944* (Guatemala: Afanes, 1996); Stefan Karlen, "Ausländische Wirtschaftsinteressen in Guatemala: Deutschland, 1871–1944. Unter besonderer Berücksichtigung der Jahre 1931–1944," *Jahrbuch für Geschichte und Staat, Wirtschaft und Gesellschaft Lateinamerikas* 31 (1994): 267–303; Stefan Karlen, "The German Colony and Economic Interests in Guatemala in the Late 19th and Early 20th Centuries," in *Grenzenlose Märkte? Die deutsch-lateinamerikanischen Wirtschaftsbeziehungen vom Zeitalter des Imperialismus bis zur Weltwirtschaftskrise*, eds. Boris Barth and Hamburger Ibero-Amerika Studien 6, (Münster: Lit Verlag, 1995), 133–156. On the history of coffee trade in Hamburg, see: Julia Laura Rischbieter, *Mikro-Ökonomie der Globalisierung: Kaffee, Kaufleute und Konsumenten im Kaiserreich 1870–1914* (Köln: Böhlau, 2011).

11 Marta Elena Casaús Arzú, *La Metamorfosis del Racismo en Guatemala* (Guatemala: Cholsamaj, 2002), 34–41; Arturo Taracena Arriola, *Etnicidad, estado y nación en Guatemala, 1808–1944*, vol. 1 (Antigua Guatemala: CIRMA, 2002), 267–389.

12 Marta Elena Casaús Arzú, *Guatemala. Linaje y racismo. Tercera ed., revisada, ampliada y actualizada* (Guatemala:F & G Editores, 2007), 67.

German immigrants soon played a central role in the Guatemalan coffee business even though they barely numbered one thousand.[13] In 1898, the German diplomat Friedrich von Erckert wrote a long report on the German presence which sheds light on their growing influence. At this point, German immigrants possessed 170 coffee fincas with a value of 64 million marks. Erckert estimated the total value of German investments in Guatemala at 184,5 million marks. Furthermore, the credits for the coffee harvests given every year by Northern German trading houses were very important for the coffee business. When Erckert published his report, the German share in Guatemalan coffee production constituted roughly one third. The region with the strongest German presence was the Alta Verapaz, where Germans owned more than 1,500 km^2 of territory and had a share of 60 percent in coffee production.[14] It was asserted, that the region was far more linked to Germany than to the Guatemalan capital. The German-Guatemalan trade agreement from 1887 gave German immigrants further privileges: They were granted total trade liberty, the right to buy land, and were guaranteed the protection of person and property. As a result, Germany became the most important market for Guatemalan coffee at the end of the 19th century and absorbed more than 50 percent of the Guatemalan production in 1890.[15]

After the First World War, a new wave of German immigrants with smaller financial resources came to Guatemala. Their hopes to acquire plantations and become wealthy were not fulfilled. During the 1920s, the German community was marked by growing social and political tensions. Nevertheless, the established Germans maintained their dominant position in the coffee business until the 1930s. The dominant position of German immigrants has provoked controversial discussions among Guatemalan historians, either portraying them as successful pioneers or crude exploiters.[16] Their success was based on the financial support from the main coffee trading centers in

13 According to official statistics, in 1893 400 Germans lived in Guatemala. As there were 900 Germans registered in the German consulates, it may be concluded that their real number was higher. See Wagner, *Los Alemanes en Guatemala 1828–1944*, 54, 325.

14 Friedrich Karl von Erckert, "Die wirtschaftlichen Interessen Deutschlands in Guatemala,“ in *Beiträge zur Kolonialpolitik und Kolonialwirtschaft* III (1901–1902): 225–284, here 225–238, 267–284.

15 Wagner, *Historia del café de Guatemala*, 103–114.

16 Wagner, *Los Alemanes en Guatemala 1828–1944*, 214–215; Julio Castellanos Cambranes, "Pioneros del desarrollo"—*¿Civilizadores? Concideraciones sobre los neocolonialistas alemanes en Guatemala, 1828–1996* , Publicaciones conmemorativas del XX aniversario 3 (Guatemala: CEUR, 1995).

Europe. In addition, they used their detailed knowledge of the European market, as well as their substantial know-how on coffee production and access to technical innovations.

Coffee Plantations in Guatemala: A Short Overview

In this section, I will give an overview on housing and technical facilities on the plantations, expand on coffee production and harvest and conclude with a short explanation of coffee processing.

By and large, the Guatemalan coffee was produced in two regions: in the Alta Verapaz and in the "Costa Cuca" area of the Western Highlands. The largest plantations were located on the West Coast, most of them owned by plantation companies founded by Hamburg merchants at the end of the 19th century.[17] The historian David McCreery named five important characteristics of the coffee plantations in Guatemala: They were large-scale, owned by foreigners, and disposed of extensive processing machinery and a diversified production. In addition, the land holdings were built up over several years.[18] He concluded:

"The Guatemalan coffee estate was a highly capitalized and technologically advanced enterprise tied to the world markets by increasingly efficient communications, but one that at the same time continued to rely on coerced labor and the direct intervention of the state for its profit and survival. This blending of modern and retrogressive systems was characteristic of colonial economies throughout the world in the last years of the nineteenth century, but in few cases was the contrast more striking than in that of Guatemala."[19]

In the Alta Verapaz, the coffee plantations were smaller than on the West Coast with a normal production between 500 and 1,000 hundredweights.[20] Still in 1940, the average area of a coffee finca in the Alta Verapaz was only

17 Katharina Trümper, *Kaffee und Kaufleute. Guatemala und der Hamburger Handel 1871–1914*, Hamburger Ibero-Amerika-Studien 7 (Hamburg: Lit Verlag, 1996), 34–47.

18 McCreery, *Rural Guatemala*, 197.

19 McCreery, *Rural Guatemala*, 195.

20 Erwin Paul Dieseldorff, *Der Kaffeebaum. Praktische Erfahrungen über seine Behandlung im nördlichen Guatemala* (Berlin: Hermann Paetel, ohne Jahr), 4. A hundredweight is a unit of mass equivalent to 50 kg.

33 acres, whereas in Western regions it varied between 68 and 107 acres.[21] Contrary to other Guatemalan regions, coffee replaced corn production around Cobán and Carcha but there was still enough land for subsistence agriculture available.[22]

As the ideal altitude for coffee growing in Guatemala was 1,000 to 1,500 meters, most plantations were located far away from cities and villages. Generally, a finca comprised the following buildings: The house of the plantation owner, a building for the higher employees, and an office where the finca books were kept. Workers' housing was normally located in separate zones. Then there were technical facilities, the processing machinery, storage rooms and patios to dry the coffee. One part of the finca was often used for the cultivation of basic grains like corn, and some land was kept in reserve, so that cultivation could immediately be expanded in case of rising coffee prices.[23]

The size and furnishings of the owner's house depended on his financial resources. Wealthy owners bought their furniture in Europe and tried to maintain a European way of living. Some of the immigrants brought their wives along with them, which led to increased social activity on the fincas. From time to time, the owners would organize social events on their plantations with meals, music and dancing. Sometimes, the women failed to adapt to the new environment and returned to Germany. Several of the German authors concurred that a coffee plantation was not an adequate place for European women.[24] On smaller or newly constructed fincas, there was frequently only a wooden house with basic facilities. Often the German owners spent several months in Europe and left their administrators in charge of the plantation.

Coffee growing took place on the surrounding fields. Coffee trees need up to three or four years until they provide the first harvest. To protect the trees from wind, cold and heat, plantation owners usually planted shade trees. They used fast growing trees, like Gravilea and banana trees.[25] A good shade tree should not attract insects and had to have small leaves and hard wood. In the best case, both trees have reciprocal positive effects on each

21 McCreery, *Rural Guatemala*, 199–200.

22 Ibid., 250.

23 Ibid., 197–199.

24 Karl Sapper, *Das nördliche Mittel-Amerika nebst einem Ausflug nach dem Hochland von Anahuac. Reisen und Studien aus den Jahren 1888–1895 von Dr. Carl Sapper* (Braunschweig, Berlin, Hamburg: Druck und Verlag von Friedrich Vieweg und Sohn, 1897), 22–23; Schmolck, *Welthandel selbst erlebt*, 93–99.

25 See Dieseldorff, *Der Kaffeebaum*, 6.

other. Sometimes, plantation owners used several variations of shade trees to prevent soil degradation showing an awareness of environmental challenges.[26]

The coffee plants are seeded in small tree nurseries located on the plantation. After a year, they are put into the soil at a distance of roughly 2x1 m for each tree. During the whole year, regular cropping is necessary to ensure a good harvest. Coffee blossom is very short and takes only two days. The trees have white flowers which have often been compared to jasmine. Coffee can be harvested six or seven months after the blossom when the coffee cherries turn red.

In Guatemala, the coffee harvest takes place between November and April depending on the height of the plantation. It was very labor-intensive as the cherries were picked by hand. Green or overripe cherries affect coffee taste negatively so that picking had to be executed carefully. The harvest workers collected the coffee in baskets carried on their back. Very often, whole families were working on the coffee plantations. The harvest workers carried the coffee from the field to the processing facilities where it was weighed and passed to processing. There was a constant lack of harvest workers and as a result plantation owners competed seriously to attract them to the fincas.[27]

To explain the method of wet processing, a closer look at the structure of a coffee bean is necessary. Under its outer skin, there is a layer of pulp surrounding the bean. Furthermore, it is protected by two other layers: the parchment and the silver skin. During the processing, the pulp and the different layers are removed. First, the cherries are put into a tank filled with water and then piped through a channel to sort out the unripe cherries and the defilement. The ripe cherries sink down and are put into a coffee pulping machine which removes the pulp by pushing the cherries through holes in a metal surface. After that, the beans are still covered by mucilage that has to be removed by fermentation. The coffee is put into large fermentation tanks where the mucilage is broken down by natural enzymes. Fermentation takes one or two days and is decisive for the coffee´s quality. A wrong fermentation can ruin a whole part of the harvest.[28] The method of wet processing required large quantities of water. Therefore, finca owners looked for water deposits nearby when choosing the place for a plantation. Wet processing led

26 Juan Antonio Alvarado, *Tratado de Caficultura Práctico* (Guatemala: Tipogr. Nacional, 1935), 47–75.

27 McCreery, *Rural Guatemala*, 228–229.

28 Wagner, *Historia del café de Guatemala*, 71–82.

to water pollution in the communities surrounding the plantations. From Costa Rica, we know that residents of the Central Valley already in the 1840s century protested against contamination. They complained about the pervasive smell but also about the lack of water in harvest times. The contaminated water caused damage to local cattle and led to public health problems. In consequence, the government approved a first, but lax legal provision in 1849. In 1901, it prohibited the finca owners from draining contaminated water into the Costa Rican rivers without additional disinfection.[29] In Guatemala and Chiapas the same problems occurred later on, when the growing demand for washed coffee led to a mechanization of coffee production in the 1880s. The finca owners also needed water for the propulsion of machinery, like the drying machines.[30]

After fermentation, coffee was dried on patios or by drying machines. On the patios, coffee needed to be raked every six hours and drying took up to ten days. On large plantations, mechanical dryers were introduced permitting a faster dehydration. The mechanization of the coffee industry increased quickly after 1880: In Guatemala, there were 686 depulpers and 230 rubbing machines in 1880. Until the 1930s, their number had increased to more than 2,700 depulpers, more than 1,000 rubbing machines, 672 polishing machines, 304 dryers and nearly 1,000 separators.[31]

After drying, coffee beans were hulled to remove the parchment and polished to take off the silver skin. Finally, the imperfect coffee beans are sorted out by women at large wooden sorting tables. They were put into coffee sacks and transported by mules or oxen to the railway stations. The coffee production and processing was frequently described by German immigrants in their reports to give the readers an idea of their daily work.

29 Andrea Montero Mora and José Aurelio Sandí Morales, "La contaminación de las aguas mieles en Costa Rica: Un conflicto de contenido ambiental (1840–1910)," in *Diálogos Revista Electrónica de Historia* 10, no. 1 (2009), 4–15.

30 Justus Fenner, "Agua y café en Centroamérica y Chiapas. Un breve recorrido histórico por la región, 1846–2011" (ponencia para su inclusión a la memoria del evento "Encuentro regional de Agua en el Sureste", Tuxtla Gutiérrez, Chiapas, Mexico, 13–14 septiembre 2001).

31 McCreery, *Rural Guatemala*, 216.

The Sources: Autobiographies and Travel Reports of German immigrants

Contrary to other European nations, Germany acquired no colonies until the late 19th century. In the context of the German nation state foundation in 1871 and the expansive foreign policy since the 1880s German emigrants were perceived as a strategic resource for political influence. This was also reflected by a terminological shift from emigrants to Germans abroad ("Auslandsdeutsche"). Whereas emigrants leave and possibly acquire other nationalities, the expression Germans abroad implies a continuous attachment to the country of origin. The second half of the 19th century witnessed a growing German immigrant literature creating a special picture of South America for the German empire as Sebastian Conrad has shown in his analysis on German immigration in Southern Brazil.[32]

In Guatemala, the first immigrants arriving in the 1840s formed part of an international merchant community. Frequently, they had already lived in several places abroad and had worked for different European merchant houses. The German state foundation changed the situation profoundly, as the next generations of immigrants referred with pride to the German Empire. At the same time, they founded their own social institutions, like the German Clubs in Guatemala City, Quetzaltenango and Cobán which permitted them to maintain a separate social sphere with exclusive character. On their coffee plantations the Germans created their own world of patriarchal rule and defended their sovereignty against local authorities as their autobiographic writing shows. In cases of conflict they either referred to their diplomatic representatives or higher Guatemalan authorities.

Several of the German plantation owners published autobiographic texts or travel reports. Their aim was to give the German audience an idea of the situation in Guatemala's coffee producing regions. Most of the texts legitimized the German presence as part of an economic and cultural "civilizing mission". Some of the texts were addressed to their own families, whereas others were written for publication in German newspapers or publishing houses. The published accounts aimed at preparing prospective emigrants for their experiences and warned of unrealistic expectations. The researcher of Romance languages and literature, Ottmar Ette, characterized travel re-

32 Sebastian Conrad, *Globalisierung und Nation im Deutschen Kaiserreich* (München: Beck, 2006), 245.

ports as "processes of linguistic and socio-cultural translation".[33] Phenomena perceived as different or strange were translated in a way that was understandable to German readers. In consequence, the texts contain frequent references to German culture. Besides, the descriptions of Guatemalan society give an idea of the writer's origins and stereotypes.

Most of the accounts follow a similar structure: First, the authors inform about their leaving from Germany and their journey to Central America. Emphasis is laid on their first encounter with the New World which frequently took place on a Caribbean island. The travelers were most impressed by the black population, which they commonly described in negative terms, whereas the indigenous population was characterized as innocent and harmless.[34] The next stage in their accounts is the arrival in Guatemala on the Caribbean Coast and the journey to the interior.

The German geographer Karl Sapper worked as an administrator on several coffee fincas in the Alta Verapaz between 1888 and 1895. During his stay, he made several long journeys through Guatemala, creating maps in the process. He wrote down his travel memoirs on the way back to Germany where they were published as a book in 1897.[35] Sapper described in detail how he passed the Rio Dulce after arriving in Guatemala in 1888.

"The way towards the interior leads through an entrance gate of really ravishing beauty. Just as the Rhine had to break through the shale mountains, the Rio Dulce had to find its way through the limestone mountains towards the sea, creating a deep valley. However, what tremendous differences in the scenery: Over there, proud castles and flourishing cities with a brave past and a vital present. The intensive traffic on the river indicates the midst of the tumultuous day. Here, there are few plantations, like pioneers of an awakening culture at the river banks, and everywhere still reins the silence of the dawning morning; over there the lines of the vines, formed in a militarily way, recount that the mature nature was subordinated by human will, here it rules in jaunty freedom and with the high spirits of youth."[36]

In Sapper's description, the coffee plantations appear as pioneers of civilizations in a wild environment. Nature in Guatemala is still young and wanton

33 Ottmar Ette, *Literatur in Bewegung. Raum und Dynamik grenzüberschreitenden Schreibens in Europa und Amerika* (Weilerswist: Velbrück Wissenschaft, 2001), 39.

34 See for example: Sapper, *Das nördliche Mittel-Amerika*, 4–6.

35 Sapper, *Das nördliche Mittel-Amerika*; Franz Termer, *Karl Theodor Sapper 1866–1945. Leben und Wirken eines deutschen Geographen und Geologen* (Leipzig: Barth, 1966).

36 Sapper, *Das nördliche Mittel-Amerika*, 10.

whereas in Germany it was already under control.[37] Sapper parted from the assumption of a linear development of human societies, emphasizing the different stages of development in the two societies. A journey to Guatemala therefore was a journey back in time, back to a stage Europe had long left behind. As I will show in the next section, most German immigrants shared Sapper's perception of tropical nature.

Descriptions of Nature and the Perception of the Tropics

The idea of a virgin, untouched forest was a part of colonial ideology creating a dichotomy between civilization and uncultivated nature. On the one hand, German immigrants portrayed tropical nature as a paradise. They described the voluptuous, rampant vegetation as a "glory of flora" outshining by far the beauty of landscapes like Lake Garda or the Riviera.[38] On the other hand, tropical nature represented a threat and a danger for them.

Helmuth Schmolck emigrated to Guatemala in 1910 and started to work as an accountant on a coffee plantation.[39] When he arrived with his suitcase at a small railway station to start his first job, he only found a message from his employer waiting for him: Schmolck was to continue on the back of a mule which was supposed to know the way to the finca. Looking back, Schmolck recalled how he perceived the virgin forest at night as suspicious, dark and silent. Arrival in the coffee zone evoked other images: Now, there were trees "lined up in orderly fashion as planted by a Prussian forest ranger".[40] The plantation signified for Schmolck the order he was used to from Germany. In a similar fashion, the protagonist in Adrian Rösch's novel sets apart the coffee plantation from the "masterless jungle".[41] Adrian Rösch emigrated from Southern Germany in 1891 and bought two coffee plantations in the Alta Verapaz, where he lived until the 1930s. Between 1913 and

37 There are parallels to perceptions of Guatemalan nature cited by Gallini, *Una historia ambiental del café*, 2–3.

38 Sapper, *Das nördliche Mittel-Amerika*, 10.

39 Two of his uncles were already living in Central America. Schmolck worked at first as an accountant on a coffee finca and then for an export firm. He published his memoirs after the Second World War when he was back in Germany.

40 Schmolck, *Welthandel selbst erlebt*, 72.

41 Oskar Weber, *Briefe eines Kaffee-Pflanzers. Zwei Jahrzehnte deutscher Arbeit in Zentral-Amerika*, Schaffsteins Grüne Bändchen 50 (Cöln am Rhein: Schaffstein, 1913), 11.

1918 he published three emigration novels under the pseudonym of Oskar Weber. His first novel, "Letters from a coffee grower" (1913) retold his own emigration history. Although the place names are fictitious, it is easy to divine that Rösch writes on Guatemala and the Alta Verapaz.[42]

The above mentioned negative perceptions of tropical nature fit in a general discourse seeing the tropics as an unsuitable place for European immigrants. In his booklet "Emigration and Acclimatization in the Tropics" (1921), Karl Sapper highlighted the tropical climate as a central danger for Europeans. According to Sapper, the "greenhouse atmosphere" of the tropical lowlands affected the work capacity of Europeans negatively.[43] Another immigrant account focused on psychological risks in their descriptions of the topography of a coffee finca in southern Chiapas:[44]

> "The residential house of Teutonia and all cultivated areas are situated in a large valley surrounded by mountain ridges. They impede the view to the vast expanse that is necessary in the tropics. After a short while, one has the impression of being in prison. After several years, a 'Tropenkoller'[45] arose."[46]

The author stressed the danger of a "Tropenkoller" because of the fears that the tropical climate could affect mental health. Another important element of this discourse was that living in the tropics also entailed on alcoholism and relationships with indigenous women. In the Anglo-American context, the same phenomenon was referred to as tropical neurasthenia. First discovered by the American neurologist George Beard in the 1860s, the concept

---

42 Elisabeth Lehnhoff, "Oskar Weber y María Schwauss: literatura e inmigración alemana en Guatemala" (Tesis de Licenciatura, Departamento de Letras y Filosofía, Universidad Rafael Landívar, Guatemala, 2004), 30–37, 50–56, 65; Weber, *Briefe eines Kaffee-Pflanzers*. The novel consists of 26 fictitious letters written between 1891 and 1913 from a German emigrant to his brother. The protagonist works as an employee on a coffee plantation until he finally buys his own land; Oskar Weber, *Der Bananenkönig: Was der Nachkomme eines verkauften Hessen in Amerika schuf*, Schaffsteins Grüne Bändchen 73, 2. Aufl. (Cöln a. Rh.: Schaffstein, 1919); Oskar Weber, *Der Zuckerbaron: Schicksale eines ehemaligen deutschen Offiziers in Südamerika*, Schaffsteins Grüne Bändchen 54, 3. Aufl. (Cöln a. Rh.: Schaffstein, 1920).

43 Karl Sapper, *Auswanderung und Tropenakklimatisation* (Würzburg: Kabitzsch & Mönnich, 1921), 45.

44 After land prices in Guatemala rose at the end of the 19th century, some German plantation owners moved on to southern Mexico where land was still more affordable.

45 The German term "Tropenkoller" can be translated as tropical madness. It refers to mental illnesses affecting Europeans living in the tropics.

46 R. A. de Brock, *Fata Morgana der Tropen oder: Tropenzauber. In 3 Bänden. Eine wahrhafte Schilderung, aber mit teilweise unwahren Namen* (Unpublished manuscript, 1934), 274.

of neurasthenia reached Britain twenty years later. Neurasthenia was defined as a mental disorder related to industrialized urban life. Its symptoms were broad, covering loss of appetite, headache, depression, and fatigue. Its transfer to the tropics was made by Charles Edward Woodruff who coming back from the Philippines published his book "The Effects of Tropical Light on White Men" in 1905. The tropical version of neurasthenia affected mainly male, white settlers in their new tropical environment. The tropical sun, the separation from the civilized world and isolation impinged on nerves and energies. The phenomenon had also a racial component as it was a white affliction as opposed to black insanity. Suffering from tropical neurasthenia was still a socially acceptable disease. The syndrome reached its climax between 1905 and 1920 but remained in medical discourse until the outbreak of the Second World War.[47] It was discussed and diagnosed in Asian and African European colonies but also in the independent Latin American states.

Whether called "Tropenkoller" or "tropical neurasthenia", the phenomenon has its historical roots in the fears of Europeans entering tropical regions. As Rebecca Earle pointed out for Latin America, Spanish colonizers feared the effect of climate and local food on their health and bodies.[48] Those fears are related to older forms of European environmental determinism as Dane Kennedy has demonstrated in his research on India. There, doctors considered the maintenance of European habits and avoiding drugs and alcohol as important therapy. Another treatment was climate change or definitive return. In Guatemala as in India, loneliness affected especially men living on isolated plantations.[49] In the 1920s the journal *The German Merchant Abroad* published several letters of a disillusioned immigrant from Guatemala. In one letter he wrote:

"The nature here is impressive, and I am indulging myself here where someone else would despair. It does not do any harm to a young person to look around in the world. If he can make a living in Germany, he should stay there. If not, he has to

47 Anna Crozier, "What Was Tropical about Tropical Neurasthenia? The Utility of the Diagnosis in the Management of British East Africa," *Journal of the History of Medicine and Allied Sciences* 64, no. 4 (2009): 518–548; Anderson Warwick, "The Trespass Speaks: White Masculinity and Colonial Breakdown," *The American Historical Review* 102, no. 5 (1997): 1343–1370; Dane Kennedy, "Diagnosing the Colonial Dilemma: Tropical Neurasthenia and the Alienated Briton," in *Decentring Empire. Britain, India and the Transcolonial World*, eds. Durba Ghosh and Dane Kennedy (London: Sangam, 2006), 157–181.

48 Rebecca Earle, ""If You Eat Their Food . . .": Diets and Bodies in Early Colonial Spanish America," *American Historical Review* 115, no.3 (2010): 688–713.

49 Kennedy, *Diagnosing the Colonial Dilemma*, 162–168.

come to terms with local circumstances, which is quite difficult for a person still having ideals. […] Here, one is stuck in the deepest jungle and has to struggle with wild Indians, toxic reptiles and health problems. Any intellectual inspiration is totally missing. In the end, many young Germans become addicted to alcohol or shack up with indigenous women."[50]

Again, we can see parallels to German colonial debates relating to the fear of adopting indigenous habits and loosing contact to the home country. In a more radical way, Carl Hagelberg[51] queried the possibilities of settlement in the tropics in a brochure dedicated to future German emigrants. He wrote:

"Though the tropics offer a wide field for our exuberant zest for action, natural laws impede our race to find a second home there. In time, it could undermine its particular character traits or it might even totally loose them. Humans can decide arbitrarily to transplant their race, but they cannot create it arbitrarily. In addition, the German planter will never be attached to his clod of earth in the same way as his brother, the German farmer at home. He will always feel as rice growing in a strange soil."[52]

Hagelberg thus negates the possibilities of German settlement in the tropics absolutely, asserting an inherent relationship between blood and soil that was used by the Nazis a decade later. But German authors also differentiated between regions more suitable for German settlement like the Verapaz highlands with its balanced temperatures.[53]

Although they were fascinated by the beauties of tropical nature, most German authors highlighted the dangers and negative effects of living in a tropical environment. For them, the coffee plantations represented islands of civilization and order in this wild context. They saw their work as part of a pioneer mission in reining in and pushing back the tropical wilderness.

The idea that it was necessary to civilize tropical nature was shared by the Guatemalan elites who incorporated it into their discourse on progress and

50 Anonymous, "Briefe aus Guatemala," in *Der deutsche Kaufmann im Auslande*, no. 1 (Januar 1924): 24–25, here 24.

51 Carl Hagelberg founded a society named "Mexikanische Siedlungsgesellschaft" and published a brochure to promote German settlement in Mexican coffee producing regions. There are some hints that he deceived several emigrants and embezzled their money. Anonymous, "Bericht des Pflanzers Walter Brösel," *Afrika-Nachrichten* Nr. 14, July 1, 1921, Deutsche Zentralbibliothek für Wirtschaftswissenschaften, Pressearchiv.

52 Carl Hagelberg, *Anleitung zum Plantagenbau im mexicanischen Tieflande* (Schleswig: Ibbeken, 1919), 6.

53 Karl Sapper, Vorwort to Adrian Rösch, *Allerlei aus der Alta Verapaz: Bilder aus dem deutschen Leben in Guatemala 1868—1930. Mit einem Vorwort von Karl Sapper* (Stuttgart: Ausland und Heimat, 1934), 6.

modernity.[54] After analyzing the general perceptions of tropical nature, I will continue with more specific aspects: I will focus on how German authors described nature and ecological aspects of the coffee plantations.

Nature on the Coffee Plantations and Ecological Considerations

The massive expansion of coffee production in Guatemala changed ecological landscapes profoundly. As I already mentioned, there is very little research on the ecology of Guatemalan coffee plantations in a historical perspective. One exception is the study of the historian Stefania Gallini who analyzed the ecological transformations at Costa Cuca related to the intense coffee production. She discusses whether the coffee fincas spawned an "ecological revolution"—a process she defines as a revolution in the material and mental construction of nature.[55] Concerning the ecological consequences, Gallini highlights two aspects: the deforestation and the effects on food production in the region. Local authorities were alarmed about the deforestation as early as the 1880s, and enacted a first protection law in 1892. Another consequence of increasing coffee production was that the local indigenous population lost its agricultural base. They had to move to higher and colder mountain territories which affected basic food production in the whole region and led to increased food imports.[56]

For the Alta Verapaz, there is no similar analysis and neither deforestation nor food production were subjects mentioned by German plantation owners, who were not in the immediate affected by them. A certain ecological awareness can be seen in four other aspects mentioned by the Germans: the adequate soil for coffee production, the use of fertilizers, the destruction of weeds and the danger of leaf diseases.

Several authors reflected on what was necessary to choose a good place for a coffee plantation. They mentioned the soil quality, the accessibility of water, the altitude and the climate. Most important for the German owners were two aspects: the yields of a plantation in the future and how long

54 Gallini, *Una historia ambiental del café*, 4.
55 Ibid., 261.
56 Ibid., 259–269.

it would take to make it productive. Adrian Rösch calculated that the first full harvest of a plantation would be possible after five years. This productive level could be held for 12 to 15 harvests, then the soil would begin to degrade.[57] Carl Hagelberg related the outputs with altitude: Plantations at higher altitudes provided good harvests for 20 years, whereas in lower altitudes the soil was already exhausted after 12 years.[58] In the Alta Verapaz, German plantation owners left the land as soon as it was not productive any more. This practice was possible because in the 19th century land still represented an abundant resource. Moreover, land in the Alta Verapaz was still cheaper than in other Guatemalan regions. Degraded soils were not considered as an important problem.

Since land rotation was still possible (if on the backs of the indigenous communities) investing in fertilizers to maintain high yields was not a priority. Even an advocate of fertilizers such as Erwin Paul Dieseldorff thought that it made only sense in times of high coffee prices.[59] In addition, he referred to the danger that fertilizer vendors cheated the plantation owners by selling useless mixtures of different substances.[60] Adrian Rösch mentioned that the mountain topography made regular fertilizing very difficult.[61] David McCreery stated as well that chemical fertilizing was not common in 19th century Guatemala because of its high transportation costs. Instead, most plantation owners used coffee pulp as fertilizer and relied on the good qualities of Guatemalan volcanic soil.[62]

When it comes to the subject of weeds, the idea of civilizing nature appeared again. Adrian Rösch's protagonist described its growth as "frightening"[63]. During the whole year, it was necessary to pull weeds off using hack and machete. If plantations remained without weeding for two years, "Bush and forest will have blanketed and ranked all our troublesome work victoriously", was the pessimistic assessment.[64] In marked contrast to this attitude, Carl Hagelberg attributed the positive effect of soil protection to weeds, but concluded that after two years, they had to be pulled up. Erwin Paul Dieseldorff agreed, arguing that weeds could as well be used as a ferti-

57 Weber, *Briefe eines Kaffee-Pflanzers*, 66.
58 Hagelberg, *Anleitung zum Plantagenbau im mexicanischen Tieflande*, 40.
59 Dieseldorff, *Der Kaffeebaum*, 21–24.
60 Alvarado, *Tratado de Caficultura*, 326.
61 Adrian Rösch, *Allerlei aus der Alta Verapaz*, 34.
62 McCreery, *Rural Guatemala*, 217.
63 Weber *Briefe eines Kaffee-Pflanzers*, 66.
64 Ibid., 66–67.

lizer if put into soil before blooming.[65] In conclusion, German finca owners had an ambivalent attitude towards weeds. They could contribute to the protection of soil and fertilizing but were seen as a danger to the plantation owners' civilizing effort as well.

During its history, coffee production was affected by fungal diseases and pathogens several times. One of the most common diseases was the coffee leaf rust discovered in Sri Lanka in the 1860s that reached nearly all coffee producing regions in the 20th century. In Central America, it did not appear until the 1970s, but coffee plantations in the 19th century were affected by other fungal diseases.[66] Erwin Paul Dieseldorff was the only author who mentioned infections of coffee trees. The most widespread disease was caused by a fungus named "Stilbum flavidum" or "ojo de gallo". It is characterized by small yellow spores on the leaves and coffee cherries which makes them fall off the trees. In his recommendations on how to combat infections, Dieseldorff draws parallels to the control of human infectious diseases. Affected trees have to be isolated soon, but he considered it even more important to prevent infection by improving ventilation on the plantations. Other infections were caused by small insects like butterfly worms and greenflies.[67]

Whereas descriptions of soil and fertilizing had mostly a neutral, scientific tone, German authors sometimes also described the plantations as a sort of forest and highlighted the beauty of the coffee blossom. Helmut Schmolck distinguished the European fields from the open tropical plantations of corn, sugar, and tobacco. He argued these plants grew up so high that it was impossible to see the surroundings when you were in it. A coffee plantation with its shade trees seemed at first like a forest, concludes Schmolck.[68] Later on, he describes a coffee plantation at blossom:

"The coffee bloom, which extends kilometers and kilometers into the forest and emits a sweet, exotic scent, can be compared with our cherry blossom or flourishing jasmine. It inspires local poets and musicians for their poetry and foreign painters for their pictures, normally only showing green branches and white blossoms. The plantation owner's pleasure in the bloom's magnificence is a more materialistic one. Seeing the blossom, he makes a still speculative calculation for the harvest. These

65 Dieseldorff, *Der Kaffeebaum*, 19.

66 McCook, *La Roya del Café en Costa Rica: epidemias, innovación y medio ambiente, 1950–1995*, 99–101.

67 Dieseldorff, *Der Kaffeebaum*, 8–11.

68 Schmolck, *Welthandel selbst erlebt*, 89.

calculations after coffee bloom influence the prices on the world market long before harvest."[69]

Schmolck's emphasis on the material interest of plantation owners parallels his perception of the coffee plantations as a "large scale industry in the countryside".[70] Natural beauty does not play a role in this. As Erwin Paul Dieseldorff stated at the end of his book, a plantation owner always had to keep in mind future surplus.[71]

For the plantation owners, nature was a resource of production that should enlarge profits. They were concerned about every global, ecological or local influence which might reduce surplus or disturb production. Coffee plantations were a new natural environment to the Germans, but also a new social environment. In the descriptions of social hierarchies, the idea of a civilizing mission can be found again. Social conflicts on the plantations are a subject generally avoided by German authors. Their main local adversary was tropical nature.

"A new world opened up for me. It is not only the scenery and the lifestyle; the social realities as well provide enough material for several volumes."[72]

Adrian Rösch's alter ego depicted his life on the coffee plantation as entering into a new world. In a detailed manner, he described the social relations on the plantation beginning with a characterization of the finca owner. Next in the social hierarchy came the protagonist himself, followed by a mayordomo, the mestizo workers and the indigenous families who had already lived there before the finca foundation.[73]

German authors generally described social hierarchies using the metaphor of a pyramid. On the top of it, there were the German employees, normally one administrator and, on large plantations an accountant as well. The next layer was formed by the "mayordomos" who were overseeing field work on the plantations. For the German administrators, they were the central contact persons with links to the foremen on the different parts of a plantation. At the lowest level of the pyramid, German authors situated the indigenous workforce, sometimes distinguishing between regular workers and harvest workers. They were characterized as an anonymous group without

69 Schmolck, *Welthandel selbst erlebt*, 90–91.

70 Ibid., 92.

71 Dieseldorff, *Der Kaffeebaum*, 36.

72 Weber, *Briefe eines Kaffee-Pflanzers*, 13.

73 Ibid., 15–17.

individual attributes. Rösch included an element of racial hierarchization in his description, distinguishing between mestizos and indigenous workers, who also lived in different areas of the finca.

Frequently, German authors used a second metaphor to describe their relations with the indigenous workers: the metaphor of a father and his children. Erwin Paul Dieseldorff formulated the following as a guideline for plantation administrators:

"The administrator must seek to build a good relationship with his workers. They should not only perceive him as a strict patron but as an advisory and helpful friend. The Indian in the Alta Verapaz should be treated as a child. To gain authority, the administrator has to be resolute and vigorous; he must have definite views and avoid changing his mind several times. On the other hand, he has to be friendly and fair-minded to gain the heart of his people."[74]

Karl Sapper shared his opinion and argued for a paternalistic treatment of plantation workers. He emphasized the need to learn the indigenous languages in order to avoid misunderstandings.[75] In his emigration guide he stated that courtesy can be omitted in the contact with indigenous people. Only tranquility and self-control are necessary.

Most of the German plantation owners had a negative perception of their workforce. They accused them of reluctance towards work, laziness and frequent robberies. In consequence, local population had to be forced or educated to work. Again, parallels exist between plantation owners' argumentation and colonial discourse, where education to work was a central part of the colonial civilizing mission. Besides, their ideas perfectly matched those of the Guatemalan elite concerning indigenous workers.[76]

As there was a constant lack of workers on the plantations[77], their recruitment was an important topic for the plantation owners. Karl Sapper saw the "desire for freedom" of the indigenous workers as an obstacle.[78] To

74 Dieseldorff, *Der Kaffeebaum*, 33.

75 Sapper, *Mittel-Amerika*, 222–223.

76 For the concept of civilizing missions in general, see Boris Barth and Jürgen Osterhammel, eds., *Zivilisierungsmissionen. Imperiale Weltverbesserung seit dem 18. Jahrhundert*, Historische Kulturwissenschaft 6 (Konstanz: UVK-Verl.-Ges., 2005); and for the German case Sebastian Conrad, *Deutsche Kolonialgeschichte* (München: Beck, 2008), 70–71.

77 The lack of workers and payments in advance was also intensely debated in Chiapas. See Sarah Washbrook, "Enganche and Exports in Chiapas, Mexico: a Comparison of Plantation Labour in the Districts of Soconusco and Palenque, 1876–1911, " *Journal of Latin American Studies* 30, no. 4 (2007): 797–825.

78 Sapper, *Das nördliche Mittel-Amerika*, 223.

avoid the government's *mandamientos*, people frequently hid in the forests or crossed the borders to avoid continuous work on the plantations. After the Guatemalan government stopped the practice, plantation owners recruited their workers by making advance payments. This practice was vindicated by Helmuth Schmolck:

> "The trick was that people always had debts. This system allowed the plantation to dispose always of workers. A worker could only leave if he received some money or was ransomed. It was a sort of peonage in disguise, perhaps necessary, because people otherwise would not work on the plantations."[79]

Adrian Rösch's protagonist argued in a letter to his brother that the low wages on the plantations made low coffee prices in Europe possible.[80] This statement represents an exception and his perception of the workers on the plantation is a more positive one: He portrayed them as "nice guys", making no trouble unless they do not appear at work.[81] Helmuth Schmolck also mentioned the low wages, but portrayed them as an advantage allowing the planters to make higher profits.[82] The historian Wade Kit concluded in his study on Erwin Paul Dieseldorff that his extensive knowledge of indigenous culture and traditions was crucial for the control of his workers and distinguished him from other plantation owners.[83] Several German administrators learned indigenous languages but in general they omitted detailed descriptions of their contact and intercultural conflicts with the workers.

Local and Global Influences on the Plantations

Many of the coffee plantations were geographically isolated places: The distances to the next cities and villages were large. Rösch described the situation on fincas in the Alta Verapaz as desperate:

79 Schmolck, *Welthandel selbst erlebt*, 78.
80 Weber, *Briefe eines Kaffee-Pflanzers*, 32.
81 Ibid., 16.
82 Schmolck, *Welthandel selbst erlebt*, 88.
83 Wade Alan Kit, "Costumbre, conflict and consenus: Kekchi-finquero discourse in the Alta Verapaz, 1880–1930" (Dissertation, Department of History, Tulane University, New Orleans, 1998), 75–97.

"However, the life on a coffee finca, especially if it is far away from a city or a village, is arduous und full of austerities. In many cases, during several months you have no company. The next neighbor lives perhaps several hours away, maybe a day's trip on bad roads. Frequently, there is little or nothing to do, which is especially bad when the weather is wet and cold. For weeks and months, one only hears the Indian language or, at best, Spanish spoken by an overseer."[84]

Communication took place by travelling and through reciprocal visits, where plantation administrators and owners exchanged business news and rumors. In the 20th century, communication technology improved and even isolated fincas could communicate by telephone or radio. Some Germans subscribed to local newspapers as a source of information, although Adrian Rösch's protagonist criticized that they only spread government attitudes and led to uncertain rumors.[85]

Local authorities were important for plantation administrators as they decided on the assignment of the workforce for the fincas. For this reason, Karl Sapper advised to be very diplomatic with local governors. However, he criticized the inefficiency and slowness of local authorities. Sometimes, he considered it necessary to pay bribes. In urgent cases, Sapper recommended the direct contact to the president.[86] Apparently, German finca owners had direct links into the circles of political power in Guatemala, which caused growing resentment among the Guatemalan population. Their dominant position in the coffee sector which increased during the coffee crisis and arrogant behavior resulted in conflicts. A Hamburg coffee firm complained in a letter to the German Foreign Office in 1902:

"Since Hamburg and Bremen trading houses had to accept in payment Guatemalan properties during the last years, a large part of Guatemalan coffee plantations is now in German hands [!]. It is probable that the Guatemalan government wants to avoid other plantations ending up in the hands of strangers. In addition, this development caused little by little a certain animosity against the German creditors, escalating in some places to xenophobia. Given the circumstances there are indications that the Guatemalan government secretly tries to prevent the acquisition of new plantations and the work on old plantations."[87]

84 Rösch, *Allerlei aus der Alta Verapaz*, 36–37.
85 Weber, *Briefe eines Kaffee-Pflanzers*, 73.
86 Sapper, *Das nördliche Mittel-Amerika*, 214.
87 Abschrift einer Eingabe der Gebrüder Oetling ans Auswärtige Amt, 17.10.1902, Senatskommission für die Reichs- und auswärtigen Angelegenheiten 1628, Staatsarchiv Hamburg.

Social unrest and changes in political power worried the plantation owners, especially during harvest times. [88] Helmuth Schmolck considered local politics as irrelevant for his business but realized the growing opposition against President Manuel Estrada Cabrera.[89]

On the global level, plantation administrators were concerned about overproduction and falling coffee prices. The expanding coffee economy experienced its first serious crisis in 1897/98. The high coffee prices during the 1880s stimulated coffee production in nearly all coffee producing countries, especially in Brazil, which resulted in an overproduction crisis. The price of Guatemalan coffee fell in one year from 32$ (1896) to 14$ per quintal (1897).[90] This coincided with a deep financial crisis of the Guatemalan government. For coffee producers it became more and more difficult to obtain credits for the harvests. Many producers had mortgaged their properties, which led to a concentration process in the coffee business. Many of the German plantation owners were on the winning side, enlarging their properties by buying up insolvent fincas. In the Alta Verapaz alone, German finca owners extended their possessions by 600 km^2.[91]

Adrian Rösch's protagonist wrote to his brother that Brazilian coffee production had been expanding for ten years. The enormous harvests were sold on the international market, so that world production exceeded world consumption.

> "As nearly all coffee came together at four points—London, Hamburg, Le Havre, New York—the consequence was a dramatic fall of prices. This was observable everywhere in the world, where coffee is produced, and it was noticeable even in the most isolated, godforsaken nook. I will have to strain everything to pay the horrible loans on my mortgages and capital."[92]

88 The protagonist of Adrian Rösch's novel described a local revolution with only few details. He was glad that the revolt happened far away and did not occur in harvest times. See Weber, *Briefe eines Kaffee-Pflanzers*, 73.

89 Schmolck, *Welthandel selbst erlebt*, 142.

90 Wagner, *Historia del café de Guatemala*, 111.

91 Ibid., 139–142; Schoonover, *Germany in Central America*, 112–136.

92 Weber, *Briefe eines Kaffee-Pflanzers*, 70.

Conclusion

The Alta Verapaz is a case where a small group of European immigrants induced a profound environmental change by introducing coffee production to the region. They implemented new agricultural techniques suited for export agriculture. Their access to capital permitted them to construct large processing facilities and apply chemical fertilizers. In general, German plantation owners were not worried about deforestation or other ecological problems arising from the expanding coffee production. Volcanic soils produced high yields intensified by the use of natural fertilizer. Chemical fertilizers were only used sparingly because they required high investment. As land was still an abundant resource, plantation owners planted new fields when soils were degraded. Nature on the coffee plantations was seen as a resource used to generate profits for the plantation owners. Their accounts reflect the optimism during the introduction of a new export crop in a period of rising world market prices.

When the Germans arrived in Guatemala, they had little knowledge of tropical agriculture and coffee cultivation. During their first years, they endured a process of learning and adaptation. Nevertheless, the German authors were convinced of the superiority of their own culture and society. On the one hand, their texts include detailed descriptions of finca organization, nature, social life, processing, technology, and coffee harvests. On the other hand, there are subjects that were completely left out such as social conflicts, the political situation in Guatemala, business failures, and intercultural misunderstandings. For the German immigrants, the coffee plantations in Guatemala were bastions of order, human dominance and progress. They considered themselves as pioneers with the task to civilize nature and local population. As a result, they constructed plantations as safe and "civilized" spaces in a tropical environment, a symbol for progress and economic recovery. However, there was an ambivalence in their descriptions: On the one hand, German immigrants were fascinated by the beauty and the variety of tropical nature and they gave detailed descriptions in their books and letters. On the other hand, the tropics represented a danger and were considered as an inadequate place for Europeans. In consequence, the authors tried to prevent illusions about living and working in the tropics. The following generations of German immigrants could count on their support but also on growing assistance by the German government which founded research institutions on tropical agriculture at the end of the 19th century.

How the education and publications by those institutions contributed to the ecological transformations of tropical regions remains an interesting topic for further research. The German immigrants shared the idea of a necessary "civilization" of tropical nature with the Guatemalan elites. Whether they engaged in the debates on deforestation and how they perceived the first protection law would need to be investigated through a more detailed revision of Guatemalan sources. This research would contribute to the larger research field of knowledge transfer, perceptions of nature and the role of German immigrants in transforming the environment of agricultural regions in the Global South.

Divide and Cultivate: Plantations, Militarism and Environment in Portuguese Timor, 1860–1975

Chris Shepherd[1] *and Andrew McWilliam*

By the time Portuguese colonialism in East Timor drew to a close in 1974 and 1975, it was estimated that 90 percent of the half-island's vegetation had been modified by man.[2] About the Baucau and Viqueque regions, 'it is hard to recognise the distribution of natural vegetation today', wrote German geographer Joachim Metzner after a period of fieldwork in 1969 and 1970.[3] Indeed, forest degradation lies at the heart of human modification of the environment in East Timor and deforestation was set to increase dramatically over the next quarter century under Indonesian control (1975–1999) at an estimated rate of 1 percent per year.[4] In this chapter we explore the relationship between the environment and the plantation as it developed under Portuguese colonial rule from 1860 until the eve of the Indonesian invasion in 1975. We focus in particular on the period from 1890 to 1940 when various crops—in particular coffee and coconut—were imposed on the indigenous population across the whole territory in the interests of economic profit. We argue that plantation cultivation was pursued at the expense of the natural environment, indigenous sovereignty, and local agro-cultural forms that worked to regulate land use and distribute benefits.[5]

---

1 For the first author, the archival research that went into this chapter was possible due to support from the Australian Research Council in the form of an Australian Postdoctoral Fellowship (DP0773307-2008-2011).

2 J. K. Metzner, *Man and environment in Eastern Timor: a geoecological analysis of the Bacau-Viqueque Area as a possible basis for regional planning*, Development Studies Centre, Monograph 8 (Canberra: The Australian National University, 1977).

3 Cited in A. McWilliam, "New Beginnings in East Timorese Forest Management," *Journal of Southeast Asian Studies* 34, no. 2 (2003): 311

4 G. A. Bouma and H. T. Kobryn, "Change in vegetation cover in East Timor. 1989–1999," *Natural Resources Forum* 28 (2004): 1–12.

5 This is not to suppose that land and resources were distributed equally. On the question of equity in development see: C. J. Shepherd, "Mobilizing Local Knowledge and Asserting Culture. The Cultural Politics of *in situ* Conservation of Agricultural Biodiversity," *Current Anthropology* 51, no. 5 (2010): 629–654; C. J. Shepherd and A. McWilliam, "Ethnography, agency, and materiality: anthropological perspectives on rice development in East

Plantation development has not figured centrally in accounts of East Timor's much-discussed environmental deterioration. In both historical and contemporary analyses, the finger is pointed at the combined effects of timber extraction (especially Timor's high quality sandalwood), overgrazing, population pressure, and traditional swidden or 'slash and burn' agriculture.[6] During the Portuguese period plantations occupied less than an estimated 3 percent of the entire land surface,[7] but plantation development logic fundamentally determined how Portuguese colonialists valued resources and how they perceived 'slash and burn' agriculture and the indigenous use of forest timbers as the principal environmental hazards. Accordingly, plantations were construed not as a problem, but as a technical solution to problems of land management; and although they occupied relatively little space in geographical terms, they were destined to cast a deep interpretive shadow over the remaining indigenous Timorese 'not-yet-plantation ecologies'.

There was a satisfying economic simplicity to the colonial logic of plantation development, behind which lay a characteristically complex politics. In this chapter, we examine the extension of plantation agriculture in Portuguese Timor as an intricate enterprise; an entangled history of crops, warfare, military occupation and forced labour expressed through the lens of environmental and cultural values. As a variant expression of that modality of development described by James Scott[8] as high modernism and its characteristic assertive prescriptions, we thus explore the military underpinnings of emergent plantation models in East Timor as well as the political, epistemic, and agricultural productivity of plantations as sites for the mutual constitution of colonial subjects as well as categories of valorized and devalorized plants, land-use and territorial power on the colonial frontier. More specifically, we

---

Timor," *East Asian Science, Technology and Society: an International Journal* 5, no. 2 (2011): 189–215.

6 H. Lains e Silva, *Timor e a Cultura do Café* (Lisbon: Ministério do Ultramar, 1956); Metzner, *Man and environment in Eastern Timor*; A. McWilliam, "Harbouring Traditions in East Timor: Marginality in a Lowland Entrepôt," *Modern Asian Studies* 41, no. 6 (2007): 1113–1143; F. J. Ormeling, *The Timor Problem. A Geographical Interpretation of an Underdeveloped Island* (Jakarta and Gravenhage: J.B. Wolters and Martinus Nijhoff, 1956).

7 Coffee alone accounted for around 2 percent of land use in the late Portuguese period, or 372 km^2 of the total 18,000 km^2. See L. M. Moreira da Silva Reis, "Timor-Leste, 1953–1975: O desenvolvimento agrícola na última fase da colonização portuguesa" (MA thesis, Lisbon: Universidade Técnica de Lisboa, 2000).

8 J. C. Scott, *Seeing Like a State: How Certain Schemes to Improve the Human Condition have Failed* (New Haven: Yale University Press, 1998).

analyse the plantation policies as a peculiar mode of colonial governance of its indigenous people that sought to maximize economic gain from available natural and human resources while minimizing concern for the environmental and cultural impacts of plantation agriculture itself. The time-worn colonial strategy of 'divide and conquer', at once highly relevant to the pacification of indigenous subjects in Portuguese Timor, can well be extended to include the plantation strategy of 'divide and cultivate' for the purpose of transforming these same subjects into productive, tax paying dependents.

The impulse to divide and cultivate hinged on a political separation of representation, where indigenous chiefs were wooed to depend on the colony for achieving personal goals but at the expense of their indigenous brothers and sisters. At an epistemic level, the concept of divide and cultivate also points to the emergence of a skewed politic that attributed differential value to certain crops, to various framings of environmental damage and benefit, and to a range of interpretations of cultivation practices. These formulations were assembled in such a way as to discredit anything that obstructed favoured plantation processes while privileging that which advanced them. Given the political censure typical of Portuguese imperialism, the persistent economic struggles of the colony, the corruption of colonial officials, and the pronounced isolation of Timor—the most remote and neglected of colonies—, this skewed politics often reached absurd proportions, permitting a level of governmental, ecological and cultural violence that the Portuguese were ultimately forced to deny before an increasingly sceptical international audience.[9] This scepticism certainly contributed to the downfall of the Portuguese republicans in 1926, but far from leading to substantial policy changes towards plantations in East Timor, state repression and censure only increased as conditions for the majority of indigenous subjects working the plantations deteriorated further.

The policy of 'divide and cultivate' as a strategy of governance was simultaneously a renewed and aggressive attempt on the part of Portuguese colonial authorities to re-order and re-configure the economic landscape of colonial East Timor. It formed an intrusive and prescriptive engagement with Timorese agricultural production where the violence of its methods flourished due to the combined effects of geographical isolation, endemic corruption, authoritarian metropolitan politics, the prevailing racial justifi-

9 W. G. Clarence-Smith, *The Third Portuguese Empire, 1825–1975: a study in economic imperialism* (Manchester, U.K.: Manchester University Press, 1985).

cations for the civilising benefits of labour and the systemic censure of opposition. For all this, however, the colonial work of plantations in Portuguese Timor failed to achieve the projected ambitious objectives. Despite the sustained efforts of successive colonial administrators to coerce and co-opt an impoverished and resentful rural Timorese population, the colonial outpost remained a drain on colonial finances and did not achieve anything approaching economic sustainability.[10]

The emergence of plantation agriculture

The first three centuries of Portuguese presence on the eastern half of the island of Timor constituted a weak form of colonialism. Until the late 19th century, the Portuguese controlled parts of the northern coast but exercised minimal influence over the upland and southern coastal areas of multiple and largely autonomous indigenous *reino* or 'kingdoms'. Relations between centers of Portuguese command and the kingdoms were characterised by reciprocities such that the kingdoms supplied foods and other goods—a tribute system designated as *finta*—in exchange for Portuguese jural authority and a measure of respect for indigenous traditions and legal codes.[11] It was under these arrangements that coffee, which was to become the colony's principal export crop, was introduced to the kingdoms during the governorship of Afonso de Castro from 1859 to 1863.[12]

Immediately prior to plantation expansion, Afonso de Castro had led negotiations with the Dutch to fix the border between Portuguese and Dutch Timor, which would serve to secure investment in coffee. Castro had also studied the forced Culture System in the Dutch East Indies where native

10 G. J. Telkamp, "The Economic Structure of an Outpost in the Outer Islands in the Indonesian Archipelago: Portuguese Timor 1850–1975," in *Between People and Statistics: Essays on Modern Indonesian History*, ed. P. Creutzberg (The Hague: Martinus Nijhof, 1979), 71–83.

11 R. Roque, *Headhunting and Colonialism: Anthropology and the Circulation of Human Skulls in the Portuguese Empire, 1870–1930* (Cambridge: Palgrave McMillan, 2010).

12 As Clarence-Smith notes, the introduction of coffee to East Timor dates to 1815 but it failed to make a significant impression for some time. See W. G. Clarence-Smith, "Planters and smallholders in Portuguese Timor in the nineteenth and twentieth centuries," *Indonesia Circle* 57 (1992): 15–30.

labourers were obliged to cultivate coffee and other crops.[13] Determined to find a substitute for the once thriving sandalwood trade and to place the colony on a more secure financial footing, Castro envisaged coffee plantations covering most of the half-island of Portuguese Timor, cultivated by the indigenous population whose numbers, he guessed, fell within the range of one and two hundred thousand souls.[14] With the creation of a *colónia militar agrícola* in mind, Afonso ambitiously carved the colony into ten (later eleven) regional military districts. But as colonial control over the territory was inconsistent, he began by promoting coffee where he could: in a handful of kingdoms with which he had negotiated vassalage. Each kingdom was supposed to develop a coffee plantation, and their rulers (*liurai*) were expected to ensure that their subjects plant 600 bushes per household on their own land. Under this plan, one fifth of the total harvest was to be handed over to the Portuguese in the form of a production tax while the remainder would be purchased by the state. These initiatives were carried on by successive governors and met with considerable success. As a consequence, large swathes of land in the districts of Liquiçá and Manatuto were cleared of forest and given over to coffee. By the early 1880s, coffee exports had skyrocketed above 2,500 tons and were already established as the principal export commodity from the colony.[15]

But as much as the Portuguese revelled in these figures, coffee was far from a mutually beneficial joint venture between the colonial state and its Timorese subjects. As coffee production expanded, so too did conflict. Local leaders (*liurai*) coveted coffee wealth as much as the Portuguese. As the latter sought to increase production, the former sought to monopolize it. Many *liurai* exploited the labor of their own subjects mercilessly just as they defaulted on the obligatory coffee tribute to the Portuguese. Where possible, they sold their coffee into Dutch Timor for better prices via Chinese merchants whose entrepreneurial flair the Portuguese were at pains to curb. As the Portuguese stepped up their efforts to control the trade of the *contrabandistas*, Timorese kingdoms rebelled. For the Portuguese, interruptions to coffee production were tantamount to declarations of war, and with mercenary forces (*moradores*) comprised mainly of Timorese soldiers from allied king-

13 G. C. Gunn, *Timor Lorosa'e: 500 years* (Macau: Livros do Oriente, 1999).

14 A. de Castro, *As Possessões Portuguezas na Oceania (1867)* (Whitefish, Montana: Kessinger Legacy Reprints/Lisbon: Imprensa Nacional, 2010 [1867]).

15 J. E. Branco and F. da Câmara, *Província de Timor: informações relativas aos jazigos de petróleo e à agricultura* (Lisbon: Ministério das Colónias, 1915).

doms, they entered enemy kingdoms and carried out pillage and head-hunting raids to force their submission to the coffee regime.[16]

Dating from the 1860s, these intermittent 'coffee wars' were punctuated by insurrections, mutinies, political intrigues, and shifting alliances.[17] By the time the infamous governor Celestino da Silva arrived in 1894, the pattern of localized power struggles was giving way to major military offensives on the part of the Portuguese authorities in order to counter escalating inter-kingdom alliances and organised indigenous resistance with weaponry purchased from the sale of coffee. Ironically, the escalation of conflict destabilised coffee production just as it was intended to drive its expansion.

Coffee, however, was also under attack from another quarter: disease. When production reached an all time high in the early 1880s, leaf rust (*Hemileia vestratrix*) set in to ravage the prized Arabica crop in the regions of lower altitude west of Dili.[18] Over the next decade, production fell steadily as coffee prices rose.[19] With the combined results of Chinese market interference, internecine conflict and the spread of coffee plant disease, Celestino da Silva encountered a coffee cultivation system in rapid decline with the majority of plantations now abandoned.[20]

Celestino, as he was widely known, was committed to turning the situation around and aimed for the complete subordination of the remaining dissident kingdoms to Portuguese rule and restoring the revenue bounty that flowed from taxes on coffee. Expertly deploying divide and rule tactics to meet his military objectives, Celestino moved to establish private plantations in pacified regions. He founded the *Sociedade Agrícola Patria e Trabalho* (SAPT) in Ermera in 1897 with leases covering several thousand hectares.[21] Military elites were the first to take advantage of the more favorable regulation for land concessions decreed in 1901 and half a dozen more such private plantations appeared over the coming years. Some plantations were established on densely forested virgin land, others carved out of fallowed forest

16 R. Pélissier, *Timor em Guerra: a conquista portuguesa, 1847–1913*, trans. Isabel Dias Amaral (Lisbon: Estampa, 2007).

17 Between 1867 and 1895 there were a dozen major revolts, culminating in the first war of Manufahi in 1894.

18 Reis, '*Timor-Leste, 1953–1975*'.

19 Cf. McCook, this volume.

20 J. A. da Silva Carvalho, "O café em Timor. Tese apresentada à Conferência Nacional do Café. Abril 1935," thesis presented to the National Conference on Coffee, April 1935 (Lisbon: Tipografia Viana, 1937).

21 Gunn, *Timor Lorosa'e*; Pélissier, *Timor em Guerra.*

areas used by the local populations for swidden cultivation, and a good many were simply acquired through the appropriation of smallholder plantations. New colonial laws were put into place to recruit indigenous laborers and force them to work the plantations at a minimum wage. On some plantations, including the SAPT, it was later alleged that indigenous captives from the wars in the kingdoms became plantation slaves.[22] Coffee predominated, but tea, rubber, cacao, coconut and other crops were also tried with mixed success.[23]

As military control over the territory was consolidated, Celestino pushed coffee onto local populations by forging strong alliances with loyal *liurai* who were eager to strengthen their grip on the coffee trade by exploiting their own subjects as unpaid labor. In 1906 a head tax (*capitação*) was instituted to replace the old *finta* system and had the effect of increasing pressure on local farmers to produce coffee and coconut for market over and beyond their ordinary subsistence needs, in order to meet the annual payment of one Mexican dollar (*pataca*) per capita. If they failed to pay, they risked imprisonment or corvée labour, building roads or tending private plantations.[24]

As Celestino was due to depart in 1908, a new production model emerged, namely, the state plantation. State plantations owed their existence to the initiative of the military commander (*tenente d'infanteria*), Antonio Leite de Magalhães, who had received training at the botanical gardens of Buitenzorg in the Dutch East Indies. He proposed that in Portuguese Timor the state should play a leading role in opening plantations across the colony. In his tract *Memoria Descriptiva dos Recursos Agrícolas da Possessão Portugueza de Timor* of 1909,[25] Magalhães sought to counter the popular metropolitan conception that Timor was too distant, its people too bellicose, its soils too marginal, its rain too inconsistent, and its geography too mountainous for profitable agricultural exploitation. Magalhães stressed the fertility of the soils, and claimed that it was the only colony where the 'absolute pacification' of the indigenous population had been achieved. The first state plan-

22 Slavery had been formally abolished in 1875.

23 J. C. Montalvão e Silva, *A Mão d'Obra em Timor. Breve memoria sobre o seu territorio, clima, producção, usos e costumes indigenas, industria, agricultura e commercio* (Lisbon: n.p, 1910).

24 Inability to pay the annual head tax resulted in forms of debt bondage for many households which contributed to the creation of an effective 'enslavement' status for those involved to patron groups which continued to influence subsequent generations of progeny. See Clarence-Smith, *The Third Portuguese Empire.*

25 A. Leite de Magalhães, *Memoria Descriptiva dos Recursos Agrícolas da Possessão Portugueza de Timor* (Dili: Imprensa Nacional, 1909).

tation was the *Granja Eduardo Marques* set up in the *comando militar* of Liquiçá[26], and within several years it would employ some 4,000 indigenous laborers (*auxilios*) and 1,000 *moradores.*[27]

When he heroically claimed the 'absolute pacification of the natives', Magalhães did not anticipate that the worst rebellion in the history of Portuguese Timor was near. Following the doubling of the head tax in 1910, the Liurai of Manufahi, Dom Boaventura, mobilized his domain and a number of allied kingdoms in a full-scale offensive against the Portuguese. Intense fighting led to the eventual defeat of Dom Boaventura in 1912 requiring a colonial army of some 12,000 troops of loyal native *moradores* to crush the rebellion. Victory for the colonial state came at a cost of around 25,000 Timorese lives[28] but the war indeed marked the end of the pacification campaigns and the definitive authority of Portuguese rule. It also marked the beginning of a new decade of feverish agricultural development.

The historic suppression of the Boaventura rebellion,[29] it was believed, would finally allow the riches of the south coast to be opened up to exploitation. To this effect, a state-military plantation based on captive labor was opened up in 1913 in southern Manufahi. The Granja 'Republica', so named to celebrate the change from monarchic to republican rule in the metropolis, was the first of a number of military plantations, particularly in those areas that were allied to Dom Boaventura and his *Manufahistas.*

The war of Manufahi taught a lesson to the Portuguese colonial administrators about the risks of excessive taxation.[30] So began a period from 1913 where officials acknowledged that taxes needed to be commensurable with the 'tributary capacity' of the tax payers themselves. A Department for the Promotion of Agriculture and Commerce (*Reparticão de Fomento Agrícola e Comericial*—henceforth the *Reparticão*) was put into operation in 1913 to assume primary responsibility for the colony's agricultural development. Military commanders were ordered to carry out the instructions of the *Reparticão.* The same commanders were also required to issue monthly reports in the *Boletim de Comercio, Agricultura e Fomento*, which was printed in Dili

26 The *granja* grew out of the experimental station of Caitocoloa for coffee and cacao. Later, it would be absorbed into the SAPT.

27 *Boletim de Comercio, Agricultura e Fomento da Provincia de Timor* 1 (1919): 114.

28 Gunn, *Timor Lorosa'e*, 177–85.

29 The Boaventura rebellion became one of the touchstones for Timorese nationalism that emerged decades later in the sunset years of colonial rule (see Araujo1975).

30 *Boletim de Comercio, Agricultura e Fomento da Provincia de Timor* 1 (1920): 62–3.

from 1914 to 1921. The singular authority of the Department and its military backing lent agricultural promotion efforts an arbitrary character and although rarely reported, violence and technical instruction went hand in hand.

But, however much Magalhães (1909) had defended the Timorese indigene as a hard worker whose occasional display of 'laziness' was nothing more than 'fatigue' due to an inadequate diet (for which he advocated higher salaries and improved conditions), everyday colonial practice told a different story. Illustratively, at the 3rd International Congress of Tropical Agronomy, convened in London in 1914, the Capitain d'Artillerie and former Governor of Timor, Jayme Augusto Vieira da Rocha, explained the necessary colonial recourse to 'force' as the inevitable outcome of the 'indigenous aversion to work':

> The Timorese native prefers not to work, he does not know how to increase production with the same amount of effort, and as he is content with the little he has… he works little or not at all unless forced to do so.[31]

Communal Plantations

With an ever-expanding network of military posts beyond the regional commandos, the *Reparticão* and the military were in a position to push for another mode of obligatory cultivation, that of the *plantacão communal*. This form of production was the invention of Governor Filomeno da Câmara (1911–1917). It was based on the original decentralized plantation model designed to prosecute war indemnities in defeated rebellious kingdoms where colonial resources were too few to directly oversee a large number of plantations in more remote areas.[32] In 1915 the communal plantation model was applied to all *comandos* across the colonial territory. Building on the new category of *propiedade communal* or 'communal property', military staff or their indigenous delegates mandated that indigenous populations construct nurseries, germinate seed, and transplant seedlings into areas presumed fit

31 J. A. Vieira da Rocha, "Organisation du travail et fournissement de la main d'œuvre dans la colonie Portugaise de Timor," in *III Congrès International d'Agronomie Tropicale*, 14 pp. (Lisbon: A Editora Lda., 1914), 6.

32 Gunn, *Timor Lorosa'e*.

for conversion for plantations. This intervention was comprehensive in scale, and was intended to penetrate every recalcitrant corner of the territory.

Communal plantations were focused on two primary crops: coffee and coconut. Between 1915–1916 in Lautem for example, the indigenous people were made to meet planting quotas for coconut (copra) production, giving rise to one million seedlings in nurseries distributed across the far eastern region of the island with injunctions for all households to plant and tend the seedlings.[33] Similarly, in the western part of the territory where coffee predominated, each family was obliged to plant 600 bushes.[34] The result of what has been termed a colony-wide 'planting fever' was that by December 1916, some 600,000 coconut seedlings and 12 million coffee bushes had been planted across the 15 *comandos*.[35]

These programs, however, were executed without attention to the suitability of soils and the vagaries of micro-climates.[36] Occasionally nursery and plantation tasks were accomplished under the direct supervision of military commanders. In general, however, there were too many plantations for this system of direct control to be implemented effectively. In Viqueque alone, for example, there were as many as 150 plantation units,[37] all implemented through a chain of delegated authority from the local *liurai,* who were now increasingly subject to the capricious authority of military district commanders. Village or *suco* chiefs acted on the instruction of the *liurai*, and village chiefs delivered their orders to local hamlet chiefs.[38] Due to an inconsistent commitment at the implementation end, communal plantations were often disorganized affairs that failed to meet the prescribed administrative targets.

In 1919, when the *Reparticão* demanded that military commanders measure and register plantations, they found themselves at a loss to do so. The dispersed 'plantations' did not correspond to any formal plantation spatialization. They were neither uniformly monocultural spaces, nor orderly laid out and clearly demarcated. Instead, many of the communal plantations were scattered amidst 'disordered' topographies and agro-forestry productive modalities that resisted the orderly expectations of the state.[39] Given this dis-

33 J. Garcez de Lencastre, "Aspectos da administração de Timor," *Boletim da Agência Geral das Colónias* 54, no. 5 (1929): 47.

34 Clarence-Smith, "Planters and smallholders in Portuguese Timor".

35 *Boletim de Comercio, Agricultura e Fomento da Provincia de Timor* 1 (1917): 82.

36 Silva, *Timor e a Cultura do Café.*

37 *Boletim de Comercio, Agricultura e Fomento da Provincia de Timor* 1 (1919): 65.

38 Gunn, *Timor Lorosa'e*, 188.

39 Scott, *Seeing like a State.*

persion, the commander of Lautem, João de Almeida, wondered how it was possible to keep track of and register the some 170,000 coconut palms in his district where plantations were little more than 'clumps of trees'.[40] Communal plantations were clearly not destined to be a great success.

The asymmetries of environmental discourse and practice

During the interwar years, environmental concerns sometimes arose in Portuguese Timor but they were highly skewed to endorse plantation cultivation and delegitimate forms of indigenous cultivation. On the one hand, there was little or no official mention of the land degradation that plantations caused or of the violence directed to landscapes and people on account of plantation promotion. As we will see in the penultimate section of this chapter, attempts to bring to light issues connected with indigenous well-being were actively suppressed. Unlike the ethical turn that fostered a new development approach in the Dutch East Indies—a change facilitated by open political debate in the Netherlands on the indigenous condition in the colony[41]—any such debate was stifled inside Portuguese Timor and in Portugal. In fact, political censure was further tightened after 1926 when the republic fell to a dictatorial regime, seven years hence to be known as the *Estado Novo*. On the other hand, the burning regimes associated with shifting or swidden agriculture among Timorese smallholders as well as their collection of forest products deemed valuable by the Portuguese, were of perennial concern to the authorities for their perceived adverse effect on forests and soils.

This marked asymmetry was inextricably linked to the way different crops were categorized and valued. In Portuguese Timor and across the Portuguese network of far-flung colonies, crops that substantially increased export earnings were referred to as *produtos ricos* (lit: *rich products*) while those of inconsequential value as export crops were assigned the lower status term of *produtos pobres* (lit: *poor products*). The rightful place for the intensified cultivation of *produtos ricos* was the plantation, whereas the term *hortas* (garden/field) was used to designate sites for the cultivation of lowly valued

40 *Boletim de Comercio, Agricultura e Fomento da Provincia de Timor* 1 (1919): 80–1.

41 See M. B. Hooker, *Adat Law in Modern Indonesia* (Kuala Lumpur: Oxford University Press, 1978), 16–17; S. Moon, *Technology and Ethnical Idealism. A History of Development in the Netherlands East Indies* (Leiden: CNWS Publications, 2007).

produtos pobres. As plantation agriculture gained pace, the Portuguese prioritized the *produtos ricos*—initially coffee, and later coconut, rubber, cacao, tea, and fibre plants such as sisal. *Produtos pobres*—especially food crops such as maize, upland rice, sweet potato, cassava, peanut, beans, and various fruits—were considered primarily subsistence crops and largely discounted. *Produtos pobres* were regarded as second order commodities because they did not fetch high prices. Reflecting the official position that came to dominate agricultural policy during the final century of Portuguese rule in Timor, one Portuguese expert observed that 'there's no justification whatsoever to grow corn, beans, rice, etc., in fields appropriate for the cultivation of coffee, cacao, rubber, quinine, etc'.[42]

By the time the *Reparticão* had come into existence, the distinction between *produtos ricos* and *produtos pobres* determined where the colony should concentrate most of its development efforts. Under this plantation-for-export logic, the *Reparticão* and its military commanders paid scant attention to the condition of the smallholder *hortas*. When they did so, it was because they recognized that in some areas where the indigenous production of *produtos ricos* had made little headway, the cultivation of *produtos pobres* could lead to a saleable surplus with which indigenous *contribüentes* could pay their taxes.[43] Authorities also came to recognize that the *produtos pobres* had importance as subsistence foods and contributed to the health of the indigenous labor force required for the production of the more lucrative crops.[44]

Despite the few reports where greater attention to *produtos pobres* was urged, the advice was not always heeded.[45] The *Reparticão*'s overwhelming interest stayed with *produtos ricos*. This resulted in what might be called an environmental blind spot where, in respect to the *produtos ricos*, only environmental problems that threatened these favored crops were accorded government attention. In no instance did the wanton destruction and burning of forest ecosystems for plantation purposes attract criticism. Environmental services and costs were simply ignored and the benefits of plantations

42 J. A. da Silva Carvalho, "Bases destinadas a promover a demarcação na Colónia, de zonas de cultura para cada espécie cultural," in *Primeira Conferência Económica do Império Colonial Português, Terceira Comissão, Colónia de Timôr*, 5 pp. (Lisbon: Ministério das Colónias, 1936), 4.

43 The *commando* of Suro serves as one example of where there was little in the way of rich-product cultivation. See *Boletim de Comercio, Agricultura e Fomento da Provincia de Timor* 1 (1919): 67.

44 Magalhães, *Memoria Descriptiva.*

45 Branco and Câmara, *Província de Timor.*

framed in purely economic terms. With a resounding emphasis on tonnage produced and revenues achieved, the plantation sector was seen as a colonial good. Illustratively, the regular issues of the *Boletim Oficial do Governo da Provincia de Timor* [46] were replete with tables detailing the performance of export crops.

In striking contrast, two indigenous activities made the colonial administrators deeply restless. The first of these practices was the indigenous collection and use of forest products, in particular the cutting of trees, which in the words of one administrator was 'a flagrant example of the vandalism to which the native submits the flora…of rich and varied species, the development of which [he] has no awareness'.[47] In 1912, a battery of legal clauses was introduced to regulate the circumstances under which the indigenous population could exploit forest products. The cutting of certain tree species now required the written authorization from the Forest Administrator of the *Repartição*, the District Chief, the Military Chiefs and the newly instituted Forestry Guard (*Guardia Floresta*).[48] Far-reaching plans to divide the territory into reserves and exploitation zones according to flora were tabled but never implemented.[49]

Secondly, administrators were greatly perturbed by indigenous farmers' annual burning of cleared vegetation that accompanied shifting cultivation cycles on forested lands given their purpose was '*just for the sake of maize*' and other *produtos pobres*.[50] Where maize was promoted in state plantation precincts, authorities insisted on the cessation of burning and the adoption of tillage carried out by the buffalo-drawn plough. Despite the wide distribution of ploughs beyond the plantations to aid farmers in the preparation of maize fields and rice paddies, the uptake was low. The commanders frequently quipped that buffalo were easier to train than the natives.[51] Neither the colony nor the metropolis stood to gain from maize cultivation and land preparation by burning. Rather, much was potentially lost. The *Reparticão*, military commanders and local administrators therefore undertook to pro-

46 This printed bulletin changed its name periodically depending on the status of the colony. The bulletins of 1838 to 1975 are available online: http://btimor.iict.pt/.

47 *Boletim de Comercio, Agricultura e Fomento da Provincia de Timor* 1 (1921): 62–3.

48 The *Regulamento Florestal*, dating from May 15, 1912, is reproduced in *Boletim de Comercio, Agricultura e Fomento da Provincia de Timor* 1 (1921): 62–3.

49 Carvalho, "A demarcação… de zonas de cultura".

50 For a Vietnamese view of the relationship between swidden regimes and rubber plantations, see Aso, this volume.

51 *Boletim de Comercio, Agricultura e Fomento da Provincia de Timor* 2 (1917: 168–71.

hibit burning practices and to impose fines on those who transgressed the prohibitions. In 1935, the persistence of burning was noted by a prominent coffee expert, Lieutenant José Carvalho, who commented that few areas were spared 'the barbaric and criminal habits of the indigenes to burn the land in preparation for their cultivation of maize'.[52] Forests and forest resources were therefore valued when measured against the indiscriminate and perceived 'criminal' burning regimes from which the colonial state could reap no direct advantage. Yet these same forests and ecologies were of no value when measured against the prospective earnings of plantations. Instead, their value lay precisely in their ruination.

Inevitably, many plantations were located within or close to secondary forest areas that were customarily used for swidden agriculture, and this too heightened anxieties that fires would encroach onto the sacrosanct plantation space.[53] In places where the risks were seen to be greatest, the Forestry Guard was authorized to monitor the area, catch offenders, and impose penalties. Despite these concerns, colonial attempts to curb swidden agriculture proved futile. The reality for most smallholder Timorese farmers was that swidden technology, including the clearing and burning of seasonal food gardens, was a vital component of their livelihoods, as were certain forms of forest product use. Attempting to prohibit the use of fire and the collection of forest timbers without offering alternative modes of subsistence was destined to fail.

Culture and the *lulik*

The environmental asymmetry discussed above was further exacerbated by a highly selective valuing of indigenous Timorese culture on the part of the Portuguese colonists. It is commonly remarked that the Portuguese were accepting of what was summed up as indigenous *estilos* (rituals), *usos* (customs), and *costumes* (traditions). The Portuguese prided themselves on their tolerance. During the nineteenth century there also arose a Portuguese liter-

52 J. A. da Silva Carvalho, "Bases destinadas a promover a íntima ligação entre as autoridades administrativas e os Serviços Agrícolas da Colónia," in *Primeira Conferência Económica do Império Colonial Português, Terceira Comissão, Colónia de Timôr*, 7 pp. (Lisbon: Ministério das Colónias, 1936), 6–7.

53 Carvalho, "A demarcação… de zonas de cultura".

ary interest and sensibility towards the *usos e costumes* of the Timorese native which, by the mid-twentieth century, had flourished into a substantial archive of books and memoria by civil servants and serving military commanders. This orientation also encouraged a distinct romanticism to enter Portuguese anthropology.[54] Portuguese acceptance of Timorese ways was so great that the pacification wars against rebel kingdoms were conducted according to the rules of indigenous head-hunting. Accordingly, when raiding and pillaging villages, the heads of enemy males would be hacked off and taken back to the victor's camp to undergo ritual treatment, while those spared decapitation (mainly younger women and children) would become domestic or plantation slaves. Controversial for many in the metropolis, colonialists in Timor accommodated these practices and fostered its expression where it coincided with their political agendas.[55] For those pragmatic Portuguese in Timor if not for their idealistic compatriots back home, when indigenous practices furthered colonial interests, the 'civilizing' part of the *civilizing mission* was practically ignored.

Other aspects of indigenous culture ran counter to colonial interests. The perceived dangers of shifting agriculture and the setting up of a *Guardia Floresta* to prevent burning has already been noted. The ceremonial slaying of female animals of a reproductive age was similarly banned as this was viewed as wasteful and a threat to livestock accumulation. The constant inter-kingdom incursions and border disputes was another vexing issue. In the colonial vision, indigenous people misspent their energies killing one another when they should have been tending to plantation cultivation and their tributary obligations. Colonial administrators and military men were thus prompted to step in as boundary arbiters to resolve local conflicts and mandate peace.

A further important feature of Timorese society that was often viewed as a persistent hindrance to the economic development of the colony was the indigenous concept of *lulik*. This is a term that expresses a combination of qualities including sacred and moral authority, danger and spiritually charged objects and places. The British Naturalist, Henry Forbes, who traveled extensively in East Timor in 1882, described the *lulik* in terms of a sacred institution, a taboo practice, something awesome and protective in times of war

54 See A. A. M. Corrêa, *Timor Português. Contribuições para o seu estudo antropológico* (Lisbon: Imprensa Nacional de Lisboa, 1944).

55 Roque, *Headhunting and Colonialism*.

and one that invoked fear of its vengeance if transgressed.[56] Even in contemporary times, when referring to land, places designated as *lulik* may not be cultivated or otherwise used without the consent of the spirit guardians (*rai na'in* or 'lord of the land') that are thought to inhabit such places and whose support must be sought through ritual invocation. Throughout Timor ritual practitioners (known variously as *lian na'in* or *dato lulik*,) bear responsibility for communicating with the *rai na'in* spirits of place, ritually appeasing them with sacrificial offerings and seeking their blessings to ward off illness and misfortune, protect growing crops, and ensure abundant harvests. The corpus of beliefs and practices surrounding the properties of *lulik* continues to inform Timorese social and religious life and extend to the contemporary existence of multiple remnant forested areas which have withstood incursion and conversion by virtue of their *lulik* status.[57]

Historically, *lulik* landscapes covered a wide variety of material forms and topographic features, but typically they included dense forests, rocky outcrops and mountaintops as well as constituent features such as springs and lakes. Behavioral restrictions applied to these spiritually charged zones, and included injunctions against clearing and agriculture, prohibitions on hunting or collection of forest products. When colonial authorities saw *lulik* country that was suited to plantation agriculture, they were loathe to see such areas excluded from exploitation simply because, in their judgment, the land was shrouded in 'native superstition'. Accordingly, indigenous people were forced to clear *lulik* lands to make way for cash cropping.[58]

It is striking how scant is the historical record in respect to how the clearing of *lulik* country proceeded in practice. Hélder Lains e Silva noted some half a century later, that 'with God knows what abuses and violence, plantations were imposed on forests whose sheer density should have engendered respect or which the autochthonous indigenous superstition consid-

56 H. O. Forbes. *A Naturalist's Wanderings in the Eastern Archipelago: a narrative of travel and exploration, from 1878 to 1883* (New York: Harper & Bros.; Oxford: Oxford University Press, 1989 [1885]), 442, 443, 446.

57 D. Hicks. *Tetum ghosts and kin: fertility and gender in East Timor* (Long Grove, IL: Waveland Press, 2004 [1976]); A. McWilliam, "Prospects for the Sacred Grove. Valuing Lulic Forests on Timor," *The Asia Pacific Journal of Anthropology* 2, no. 2 (2001): 89–113; McWilliam, "New Beginnings"; E. G. Traube, *Cosmology and Social Life: Ritual Exchange among the Mambai of East Timor* (Chicago and London: University of Chicago Press, 1986).

58 Clarence-Smith, "Planters and smallholders in Portuguese Timor"; Silva, *Timor e a Cultura do Café*.

ered sacred'.[59] The following testimony offered by Júlio Garcez de Lencastre (a military commander posted to Lautem in the mid-1910s) and registered by Pinto Correia provides a glimpse into the direct dealings of Governor Filomeno da Câmara with the *régulo* of Ossu in the southeast of the country, where in 1916 *lulik* lands were destroyed to make way for the advance of 'communal plantations'.

The *indígenas* of Ossu consider sacred the great forest that covers one of their mountains, one that even today is known by the name *Mundo Perdido* that the enthusiastic agriculturalist, Filomeno da Câmara called it. When the Governor was traversing these lands for the first time, he was reminded of the *Lost World* of Conan Doyle, and he immediately decided to convert the beautiful forest into an enormous and lucrative coffee plantation. He addressed the mountain inhabitants and drew their interest, bringing to their imagination marvellous and dazzling images of riches. When he returned, the *indígenas* had already undertaken the numerous and complicated *estilos* [rituals] required to placate the *lulik* spirits, who soon saw themselves dispossessed of their dominions.

Filomeno was welcomed by the civil [elected] chiefs, the religious leaders of Ossu, the traditional elders and the *rai-lulics* [ritual practitioners], accompanied by the entire population. The men were preparing to clear the forest and the women were beating the festival drums. And then the corpulent figure of *régulo* Dom Francisco of Ossu appeared and his voice, brusk and authoritative, addressed the Governor: "*Amo* and *Senhor*, we, the Timorese, no longer need weapons because never again will we make war. What we need is manpower. Your excellency, accept as our pledge of peace this gift, the most *lulik* of our forebears." And upon his word, he handed Filomeno da Câmara a fantastic rifle (*espingardão*), whose barrel measured two and a half meters, the most prestigious relic that had been kept inside the most important *uma lulic*[60] of the *lulic* forests of Ossu.[61]

In effect, Timorese chiefs sacrificed their forests in order to make peace with the Europeans. They did so in the name of communal plantations, which, as it turned out, were not particularly communal at all.

59 Silva, *Timor e a Cultura do Café*, 41.

60 The *uma lulik*, or 'sacred ancestral house' represents a ritual and ceremonial centre for the agnatic members of a clan group who periodically convene at the *uma lulik* for life cycle rituals and invocatory prayers to the ancestors. The *uma lulik* also serves as a repository for ancestral heirlooms and objects of veneration.

61 A. P. Correia, *Timor de lés a lés* (Lisbon: Agência Geral das Colónias, 1944), 129–30.

Divide and cultivate

Forced cultivation induced indigenous subjects to plant the designated quantities of seedlings with the least possible effort, resulting in 'poor quality, overly dense, humid, dark, unmaintained, and unproductive plantations'.[62] In stark contrast, other plantations were formed by burning lands including the felling of the largest of trees, giving rise to plantation landscapes devoid of shade trees leaving the coffee trees over-exposed to the sun.[63] Many communal plantations were quickly abandoned while others were set on fire and destroyed in protest.

The existence of leaf rust in Timor rarely receives mention in the some 90 issues of the *Boletim de Comercio, Agricultura e Fomento* (1914 to 1921), and in 1929 it was reported that 'Hemilea vastratix and other diseases, that have occasioned so much damage in Java and Ceylon, have still not invaded the plantations of Timor'.[64] The first substantial reference to the prevalence of Hemilea in Portuguese Timor appears in Carvalho's 1935 thesis *O Café em Timor*.[65] In retrospect, this disease may account for the tendency of the Portuguese to abandon rather than replace ageing plantations in favour of developing virgin lands, thereby capitalizing on the 'forest rent', as the Dutch had done in Java half a century earlier.[66] It may also explain the enthusiastic introduction of Liberian coffee in the late 19th century.[67] For José Carvalho and Hélder Lains e Silva, the non-recognition of leaf rust was one more example of colonial short-sightedness and technical incompetence, but it may also be linked to the persistent but generally unsuccessful attempts to draw European settlers and investors to Portuguese Timor.[68] The sharp decline of coffee tonnage havested in Portuguese Timor in the 1880s at a time of record high prices, however, would seem to confirm the prevalence of leaf rust.

With or without leaf rust, yields of the communal coffee plantations were famously low or non-existent. For both coffee and coconut, critics accused

62 Silva, *Timor e a Cultura do Café*, 41.

63 Carvalho, "O café em Timor," 14–15.

64 Lencastre, "*Aspectos da administração de Timor*," 24.

65 Carvalho, "O café em Timor," 15.

66 Leaf rust became the major focus of coffee research in Portuguese Timor in the postwar period, particularly following the discovery of 'The Timor Hybrid', a resistant Arabica variety found amidst a plantation of affected trees.

67 See McCook, this volume.

68 See Magalhães, *Memoria Descriptiva*; J. dos Santos Vaquinhas, *Boletim da Sociedade de Geographia de Lisboa* 7 (1887): 453–61.

the government of wasting money on communal plantations that failed to yield anything.[69] Eventually it was admitted that these plantations had been a failure. Carvalho notes that 'communal plantations were received with… distrust' and that 'in the majority of cases, the "communal plantation" only served to oblige those who would receive no benefit whatsoever to look after them, while the chiefs and half a dozen elders divided the harvest amongst themselves.'[70]

The domination of chiefs over commoners in Timorese society is vividly illustrated in the case of the plantations of the mountainous region of *Mundo Perdido*. In 1928, some 12 years after the forest had been turned over to coffee, the administrator of Baucau, Pinto Correia, visited Ossu where he encountered Dom Francisco 'already old, limping with rheumatism, incapable of riding a horse' yet who 'in the following year acquired a *Chevrolet* automobile for his personal use, and a truck for the transport of goods to and from the region.'[71]

Dom Francisco of Ossu was one of a generation of chiefs across Portuguese Timor who prospered through privileged engagement with the Portuguese authorities. As Clarence-Smith (1992) notes, communal plantations, where successful, were generally appropriated by chiefs who had more to gain by extending their loyalty to the Portuguese.[72]

According to Carvalho, there were about a dozen major indigenous chiefs who ran large coffee plantation estates on private property and whose success was augmented by the policies of Governor Teófilo Duarte (1926–1929).[73] As severe economic depression was looming in Portugal, one last coffee, rubber and coconut planting fever was resumed with fanaticism by Duarte.[74] By this time, indigenous chiefs (*liurai*) were the main beneficiaries, while many commoners lost their lands and were made to work for free. Gone was the idea that the Timorese commoner could be redeemed. The attitude of incoming govenor Maj (Dr Paul Manso Preto) towards the Timorese as 'lazy, perfidious, boastful, stuck in their ways and customs' highlights the low

69 *Boletim de Comercio, Agricultura e Fomento da Provincia de Timor* 1 (1921): 52.

70 Carvalho, "A íntima ligação entre as autoridades administrativas," 14.

71 Correia, *Timor de lés a lés*, 130.

72 C. J. Shepherd and A. McWilliam, "Cultivating plantations and subjects in East Timor: a genealogy," *Bijdragen tot de Taal-, Land- en Volkenkunde* 69 (2013), 326–361.

73 Carvalho, "O café em Timor," 5.

74 Clarence-Smith, "Planters and smallholders in Portuguese Timor".

regard in which they were held.[75] In a context where land concessions were granted only to a small number of '*asimilados*', '*civilizados*' and indigenous chiefs, the governor's solution to 'the native problem' was that they simply be compelled to work for the financial benefit of the state.

Given the prevalence of purified colonial accounts and the absence of indigenous Timorese voices, the level of destruction to land and people under this authoritarian regime can only be imagined. Whether plantations were communal or the private holdings of chiefs, many that appeared to have initial success soon went to ruin. *Mundo Perdido* was one of the many coffee plantations situated in the eastern part of the territory that turned out to be a dismal failure. In 1936, Lieutenant José Carvalho lamented that the famous plantation of *Mundo Perdido* in Viqueque 'where the natives were obliged to cultivate… more than a million coffee bushes… was carried out without any prior technical study'. If initially 'the plants thrived wonderfully, when the roots reached the calcareous layer, from one year to the next could be witnessed the withering and death of tens of thousands of plants'. 'All over the island', regretted Carvalho, 'there were countless such cases'.[76]

Market, labor, and prison plantations

Agricultural policy in Portuguese Timor was conditioned by market forces. The sharp decline of coffee plantations towards the end of the 1910s can be attributed to the collapse of the market during World War I. Their revival in 1927 also followed a surge, albeit brief, in the international coffee trade. Indigenous chiefs were also attentive to market prices, picking the fruit when it seemed worthwhile, and leaving it when it did not. As we have seen, the colonial authorities came to rely on indigenous chiefs for production, and the majority of technical interventions and incentives were delivered to this indigenous elite. The knowledge that this elite relied heavily on forced labor or even forms of indigenous slavery was of no concern to most Portuguese in Timor.

75 Raúl de Antas Manso Preto Mendes Cruz, *Boletim Geral das Colónias* 97, no. 9 (1933): 276–302.

76 Carvalho, "A demarcação… de zonas de cultura," 4.

At the turn of the twentieth century a legal code was put in place throughout the empire defining indigenous labor as 'a legal and moral obligation'. In theory, although labour was involuntary, a host of clauses set out the employers' obligations in terms of salaries and conditions. However, for historian Clarence-Smith, 'the gap between legislation passed in Lisbon and practice on the ground was nowhere so vast as in the field of labor'.[77] Some private plantations on Timor dating to 1897 availed themselves of slave labor in the form of captives from the pacification wars[78], but state plantations drew on the forced but remunerated labor of indigenous farm workers or *auxilios* under contract with indigenous chiefs who charged a fee for their intermediary service. On the plantations, working conditions were hard, wages were miniscule, and harsh punishments were dealt to those who slackened off or fled. Worker conditions deteriorated further in 1921 with a policy shift whereby government plantations would become the basis of a *Colonia Penal Agrícola*. Henceforth, the government would rely less on salaried labor, instead drawing on a surplus of prisoners including indigenous Timorese who failed to pay their head tax and *deportados* arriving from Portugal, Macau, and the African colonies.[79]

There are few written records on these penal plantations on Timor. Portugal had for some time attracted severe international criticism for its treatment of indigenous populations.[80] Portugal's right wing swept aside the republican government in 1926, which had been much discredited internationally and at home by the information and rumors that circulated on the colonial labor process. What was soon to become the Salazar dictatorship was keen to inspire international confidence in its ability to be more humane in colonial management. Yet reforms were only on paper, while repressive methods were scaled up amidst tightening censorship.[81] Significantly, Portugal refused to become party to ILO conventions, while in the colonies the arbitrary exercise of power accompanied a flourishing climate of corruption. Its distance from the metropolis and its relative political and economic irrelevance meant that Portuguese Timor was to stand out in these respects.

Under the right wing governance, what *could* be known about Timor was highly contrived, while the conditions under which indigenous and other

77 Clarence-Smith, *The Third Portuguese Empire*, 139.
78 Gunn, *Timor Lorosa'e*, 216.
79 *Boletim de Fomento de Timor Portugues* 1 (1921): 54–5.
80 Clarence-Smith, *The Third Portuguese Empire*, 140–1.
81 Ibid.,180.

prisoners worked on government and prison plantations only came to light through tales recounted by visitors. In 1929 two articles published in Dutch Indies newspapers reported the experience of a certain Dutch-Javanese plantation expert, Conrad Hendrik Ryshouwer, who had been contracted in 1927 by Portuguese Timor to assist with the development of a rubber plantation in Viqueque, where some 1,200 prisoners travailed at a time when rubber was fetching high prices. The 'reign of terror' led by the *Repartição* and implemented by military commanders was recounted in graphic detail; cruelty, the arbitrary recourse to violence, the high incidence of disease and mortality rates, and the absence of shelter and food were among the allegations.[82] The Consul of Portugal in Surabaya (East Java) relayed the articles to the Ministry of Foreign Affairs in Lisbon, who then demanded an official enquiry into the case. The resulting investigation refuted the claims, and impugned the integrity of the rubber expert by raising questions about his racial stock, his communist sympathies, and his level of expertise on rubber. Ryshouwer had, it was finally alleged, become embroiled in business that lay outside the technical terms of his contract.[83]

In 1932 an article published in the Australian *Smith's Weekly*, based on two informants, confirmed the 1929 reports published in the East Indies newspapers, and elaborated on the cruelty dealt to prisoners who for their 'inability to pay their taxes, are put to forced labor in chains… and if they weaken in their tasks are flogged with long bamboo rods until they fall exhausted and bleeding…'.[84] The system of *Colonia Penal Agrícola* was further institutionalized in the 1930s, and at the end of that decade Portuguese Timor was classified as 'a penal colony' as captive labor from Macau and Goa was increased.[85]

82 "Nuttiger Werk voor den Volkenbond. Uit Duister Portugeesh Timor. Slavern .. der Inheemschen…lfstraffen en Mishandelingem," *Surabaya Courant*, August 15, 1929. A similar article appeared in *Java-Bode* on July 12, 1929.

83 Colónia de Timor, *Inquerito sobre o incidente entre o comandante militar de Viqueque e um pratico Javeanez ocorrido em fevereiro de 1929*, Dili, March 1930 (Arquivo Histórico Ultramarino 1724-I 1c MU DGCOr Mc 1928).

84 Cited in Gunn, *Timor Lorosa'e*, 212.

85 See Gunn, *Timor Lorosa'e*, 212–13.

Postwar reflections

Portuguese Timor was devastated as a result of Japanese occupation, and as many as 60,000 Timorese perished, falling victim to starvation and disease or Japanese retribution for the support the Timorese offered to the allies.[86] As the Portuguese assumed repossession of Timor in 1945, an immediate priority was to revive the plantations that had become overgrown; funds were quickly dispensed from Portugal to assist European plantation owners.[87] For the most part, the structure of plantation agriculture that had emerged before the war was reinstated, and a surge of Portuguese private investment into plantations commenced in the early 1950s. Forced and prison labor continued as before, and chiefs resumed production on their large landholdings.[88] José Ramos-Horta, destined to become the President of postcolonial East Timor decades later, described one chief of Atsabe as a despot for whom 'thousands of men and women were forced to pick [his] coffee every year and carry it on their backs to storage miles away…'.[89] The business of such chiefs was not only supported by the agricultural services, they were protected by the secret police—known as PIDE—whose powers were extended following a rebellion in Viqueque in 1959.[90]

Smallholder production of coffee independent of chiefs also increased under an incipient postwar development ethic that stated that Portuguese should 'interest' indigenous farmers in cash crop cultivation with incentives and rewards rather than oblige them to it.[91] It is not clear, however, to what extent forced cultivation diminished as more voluntarist forms of agricultural promotion were instituted. Certainly, discourse continued to conceal the abuses and exaggerate the benevolence of the colonial authorities, while the level of force deployed depended on the whims of district administrators.[92]

In the interwar years, the little technical expertise directed towards plantation agriculture had largely fallen on deaf military ears. In the postwar

86 Gunn, *Timor Lorosa'e*.

87 Reis, "Timor-Leste, 1953–1975".

88 On prison labor in the postwar period see J. Joliffe, *East Timor: nationalism and colonialism* (St. Lucia, Qld: University of Queensland Press, 1978).

89 J. Ramos-Horta, *Funu: the unfinished saga of East Timor*, preface by Noam Chomsky (Trenton, NJ: Red Sea Press, 1987), 30.

90 Gunn, *Timor Lorosa'e*.

91 Reis, "Timor-Leste, 1953–1975"; H. Lains e Silva, "Programa de desenvolvimento agrícola 1965–1975," *Comunicação* 47 (Lisbon: MEAU, 1964).

92 Metzner, *Man and Environment in Eastern Timor*.

years, in contrast, the status of technical knowledge changed dramatically. When the interwar plantation culture came to be increasingly viewed in light of the postwar advances in technical knowledge, Hélder Lains e Silva—the afore-cited agronomist sent to Timor by the *Junta de Exportação do Café* to assess the coffee industry—came to exert considerable influence on the direction of agricultural promotion. Indeed, scientific advances in botany, climatology, and the soil sciences (pedology and edaphology) were now being applied to Portuguese Timor by Lains e Silva and Ruy Cinatti, the latter being the head of the Agricultural Services in Timor during the early postwar years. Together, these men classified the soils and climate regime across the colony, and superimposed soil maps, botanical maps, and hypsometric charts to produce recommendations of where, exactly, to promote which varieties of coffee.[93]

In addition, for the first time plantations were targeted as sites of environmental degradation and new protection measures were implemented. Positioning the deterioration of forest as Timor's most pressing environmental problem, agro-forestry projects were undertaken to bring plantation cultivation into line with erosion prevention methods by promoting a switch from the use of the shade tree Casuarina to that of *Albizzia moluccana* and similar leguminous trees; this method promised to rejuvenate forest cover and protect soils.[94] These interventions were, however, limited to a few regions, and it was reported that Ruy Cinatti had little success convincing the indigenous leaders of the need for conservation.[95] Somewhat ironically, science was also prompting a critical re-evaluation of the value of indigenous culture, the preserves of superstition and rationality, and the meaning of *lulik*. Remarked Lains e Silva in 1956, 'the "irrational" beliefs of the Timorese… that bring them to religiously respect *lulic* forests, constitute a lesson from which we should all learn.'[96] Lains e Silva's endorsement of indigenous culture was but a brief interlude, deployed perhaps to attack some of his contemporaries who continued to apply the agricultural methods of temperate climates to tropical ones, rather than to seriously consider a return to *lulik* precepts. With greater earnestness, Lains e Silva went on to explain the indigenous logic with which indigenous subjects appropriated coffee cultivation in such a way as to minimize labor input; the discrepancies (in weeding, spacing,

93 Silva, *Timor e a Cultura do Café*, 86, 146–9

94 Ibid.

95 Ibid.

96 Ibid., 92.

yields etc.) between European and indigenous plantations that José Carvalho had observed earlier, Lains e Silva reframed as a rational outcome of selective indigenous ecological adaption of European crops.

Yet this was certainly no uniform vindication of peasant rationalities as would emerge in other tropics.[97] It should come as no great surprise that on the eve of the green revolution, Lains e Silva went on to propose a strategy to reduce indigenous maize cultivation (and its associated 'destruction') in the uplands by two thirds in favor of coastal wet-rice development. Those who remained in the uplands, Lains e Silva insisted, should do so only in dedication to coffee cultivation.[98] With the government's efforts to turn uplanders into lowlanders through south coast resettlement schemes, the agricultural failures and the physical sickness (e.g. malaria) that the indigenous people encountered there, provided them with ample evidence that local *lulik* spirits were resisting their presence in the land. For the most part, they returned to the uplands, reinstated ancestral rituals and resumed agriculture-as-usual.

As the departure of the Portuguese became imminent, a band of several hundred mestizos and Timorese *civilizados* turned to politics. Amidst a last minute flurry of nationalism and political party formation, two parties emerged to represent plantation elites, while another, Marxist-oriented party promoted an agrarian reform and began forming agricultural cooperatives (including for the production and marketing of coffee), the democratic likes of which Timor had never seen. Indonesian pressures (behind which lay cold-war geopolitics) and escalating internal divisions, however, proved too much for the successful transfer of power from the Portuguese to the Timorese.[99] As the Portuguese withdrew, Indonesia invaded, and so plantation agriculture was poised to take yet another turn under a different regime that devised its own methods—even more violent than those of the Portuguese—for extracting the most possible from Timorese soils and Timorese peoples while attempting to integrate East Timor into the Indonesian nation.[100]

97 Cf. Shepherd, *Mobilizing Local Knowledge*.

98 Silva, "Programa de desenvolvimento".

99 J. Dunn, *Timor: a people betrayed* (Sydney: ABC Books, 1996 [1983]); Joliffe, *East Timor*; Ramos-Horta, *Funu*.

100 See Shepherd and McWilliam, *Cultivating Plantations*.

Conclusion

In this paper we have shown how the chronic inability of Portuguese Timor to generate profitable returns for its colonial rulers prompted a series of interventions beginning in the 1860s to promote plantation agriculture as a prospective solution to continued economic malaise. Coffee and coconut were the favored crops whose propagation was extended throughout the country via a combination of the co-option of indigenous rulers (*liurai* and *dato*), coercive policies towards vanquished rebel kingdoms, and a sustained orientation towards the use of forced labor (paid and unpaid) and extractive taxation regimes.

The development of the plantation sector was privileged over the perceived destructive and low-value swidden agriculture of Timorese farmers. Colonial attempts to eradicate the clearing and burning of seasonal food gardens, while ultimately a futile gesture, nevertheless contrasted with the alacrity by which forest reserves were clear-felled for plantations with scant regard for environmental impact, land degradation or even the suitability of the locations for approved plantation crops. The developmentalist logic of plantation agriculture informed successive generations of colonial land management policy leading to a variety of experimental approaches including privatized, state and communal forms of plantation development. In the end however, the picture that emerges from the multiple efforts of successive Portuguese governors to promote commercial plantations is one of continued frustration and failure to meet production targets.

This sustained enthusiasm of the colonial state in Portuguese Timor for a prescriptive industrial approach to plantation forestry, despite its numerous setbacks, has many of the hallmarks of what James Scott has described in terms of authoritarian high modernism. The impulse of the nation state for the 'administrative ordering of nature and society', for legibility and simplification, leads, Scott has argued, to a sweeping vision of the manifold benefits to be derived from the rational engineering of social and economic life. This is a vision sustained by

> a supreme self-confidence about continued linear progress, the development of scientific and technical knowledge, the expansion of production, the rational design of social order, the growing satisfaction of human needs and, not least, an increasing control over nature... .[101]

101 Scott, *Seeing like a State*, 89.

But this ebullient confidence of high modernist visions, as Scott has demonstrated in forensic detail, is frequently found wanting and continually frustrated by its own *hubris* and simplifying fictions. As he notes, in a commentary that mirrors the disdain of Portuguese administrators towards their Timorese farmer subjects, '(T)he logical companion to a complete faith in the quasi-industrial model of high-modernist agriculture was often an explicit contempt for the practices of actual cultivators and what might be learned from them'.[102] The fact that colonial authorities in Portuguese Timor found themselves relying on the use of prisoners as laborers on plantations by the close of the 1930s, underscored both the economic and moral bankruptcy of the colonial plantation sector and, in many respects, the failure of the entire colonial project which, in the case of Portuguese Timor, always held out the prospects for greater bounty than it ever delivered.

102 Scott, *Seeing like a State*, 304.

Pines, Pests and Fires: Large Scale Plantation Forestry in New Zealand, 1897–1955

Michael Roche

> Plantations had a long history in the conquest of tropical regions by European powers and agencies from the sixteenth century onwards, and, until the abolition of the slave trade, operated on the basis of the most heinous appropriation and cruel use of slave labourers. The reintroduction of plantation agriculture into many European colonial territories in the nineteenth and twentieth centuries during the gradual eradication of slavery was a continuation, in slightly different form, of the exploitation of poor labourers. It was also symptomatic of the exposure of the world outside the metropolitan hearths of Europe to change to a more fully fledged advance of global capitalist relations of production… The management and finance of these plantations was usually though not exclusively in the hands of European or North American enterprises, and normally reflected actual or perceived demands from industrialised countries of the world for food or industrial raw materials'.[1]

Sugarcane, cotton, and slavery occupy a central place in plantation agriculture historiography. But this does not encompass the sum total of plantation systems nor one of its more overlooked forms, plantation forestry, which is arguably one of considerable contemporary importance in the light of concerns over monoculture vulnerability and loss of biodiversity. Plantation forestry may be thought of as a part of a system of 'scientific forestry' that developed particularly in Germany and France from the 17th century and was spread throughout European overseas empires. In remote parts of the British Empire such as New Zealand, exotic plantation forestry, rather than sustained yield management of natural forests came to dominate state and corporate forestry effort and expanded rapidly and on a prodigious scale.

In New Zealand, located in the temperate mid-latitudes of the Southern Hemisphere, the timing of the first major plantings in the 1920s and 1930s, as well as the conditions under and purposes for which they were established by the state and private sector, for local needs and not as industrial raw material for Northern industrial concerns further disconnects plantation

1 Robin Butlin, *Geographies of Empire* (Cambridge: University Press, 2009), 543–544.

forestry from the plantation agriculture systems discussed by Butlin. But for all that, to foreshadow the argument developed later in the chapter, there are still a number of aspects of the plantation forestry story in New Zealand where labour availability and conditions, access to land, and the raising of capital connects exotic plantation forestry to wider discussions about plantation systems. Butlin's summary of plantation agriculture does, however, omit overt reference to environmental risk, which is discussed in this chapter.

The first part of the chapter provides a brief overview of deforestation in New Zealand as the driver of exotic plantation forestry. This is followed by a section that examines the early state response from 1897 to 1919. After this, the main body of the chapter discusses the large scale state planting and company plantings that took place in adjacent districts during the 1920s and 1930s.[2] Finally the risks posed by fire and insects to plantation forests are briefly considered.[3]

Deforestation and Imminent Timber Famine

Scientific forest management gained traction from the 18th century by alleviating fears that various European states were no longer self-sufficient in timber. Rhetoric about an impending 'timber famine' was also a feature in many areas of 19th century European expansion overseas. New Zealand was no exception. Its evergreen indigenous forests comprised coniferous-broadleaf species and a southern beech forest (*Nothofagus* spp.) with the latter concentrated on the mountainous lands of the South Island. Many of the forest species were highly endemic and close to 80 percent of the country was forested when Polynesians first settled the country around 1250–1300 AD. This proportion had fallen to 54 percent by 1840 when the country became part of the British Empire. By 1900, after 60 years of colonial settlement, the forest had been reduced to around 25 percent of land area largely by burning to clear land for agricultural and pastoral expansion, although the timber in-

2 This article concentrates on the era up to and including the 'first planting boom' of 1925 to 1934 and follows these forests through to maturity but does not discuss the subsequent second planting boom of the 1960s to 1980s or the third phase of the 1990s.

3 Fungoids also posed a threat but proved to be of comparatively lesser significance especially in the 1897 to 1955 period and have been omitted here for reasons of space.

dustry had contributed significantly to the diminuation of the Kauri forests (*Agathis australis*) of northern New Zealand.

Efforts to introduce European style forest management practices date from 1876 when Captain Inches Campbell Walker, a British forestry officer from Madras was appointed as Conservator of Forests, but with weak legislation he departed after little more than a year.[4] By the 1890s however, the belief in inexhaustible supplies had ended; the forest was no longer regarded as a barrier to settlement but was instead seen as threatened and from 1894 to 1905 some 2.75 million acres [1.1 million ha] in remote areas were set aside in order to permanently preserve the indigenous flora and fauna for aesthetic, scenic, and scientific purposes in six National Parks along with 40,500 acres [16408ha] in small scenic reserves from 1892 to 1910. Another 1.29 million acres [5,220sq km] of forest on Crown Land was gazetted as Forest Reserves of various types, from 1890 to 1919, to be felled at a later date.[5]

The Timber Conference of 1896, a meeting of saw millers and government representatives affirmed a coming timber famine. The state's response was to create a Forests Branch of the Lands Department in 1897. Its officials agreed with the accepted wisdom that the merchantable indigenous forests were too slow growing and difficult to propagate to be saved in the long term and that the remaining supplies ought to be eked out in an efficient manner, with long term future supplies to be met from yet to be established exotic plantation forests. This viewpoint was to some extent also underlain by a particular interpretation of J.D. Hooker's ideas about the 'displacement' of species, which presented the New Zealand flora as 'weaker' than invading northern hemisphere species and falling back before their advance. The fatalistic extrapolation was that the indigenous forests could not easily be protected in the long term and that the only option was to harvest them fully in the immediate future.

The decision to trust in exotics for future timber needs also rested on the success that had been encountered since the 1850s in introducing a range of familiar European and other new North American and Australian forest tree species to New Zealand, where many grew much faster than in their home environments and certainly much more quickly than the main in-

4 Lanna Brown and Alan McKinnon, *Captain Inches Campbell Walker, New Zealand's first Conservator of Forests*, New Zealand Forest Service Information Series 54 (Wellington: Government Printer, 1966).

5 *Appendicies to the Journals of the House of Representatives* (hereafter *AJHR*) C6 (1906), *AJHR* C6 (1910) and complied from the *New Zealand Gazette*, 1890 to 1919.

digenous timber trees. For instance a merchantable Douglas fir (*Pseudosuga menziessii*) matures in about 45 years compared to Rimu (*Dacrydium cupressinum*) in 200 to 300 years. In the absence of professionally trained foresters who might have advocated sustained yield management of natural forest, the farming community, virtually unopposed till the 1890s and the timber industry continued to make special claims for continued cheap access to standing forest. This made it difficult for officials to secure even 'efficient' harvesting of the forests as a one off crop from the land. Plantation forestry with fast growing exotics seemed to offer the only long term solution.

The sequence of events in New Zealand of identification of a timber famine followed by government intervention to regulate forest use and create a new plantation forest estate broadly follows that of other European and settler societies. It is distinguished however, by its timing, comparatively late in taking place, the central role of the state, which pragmatically filled some spaces occupied by private enterprise in Europe, and importantly by the commitment to a plantation forestry initiative based on very fast growing though, in the case of *Pinus radiate,* unproven exotic forest species.

The Forestry Branch and the Royal Commission on Forestry 1913

The small Forestry Branch despite its name was concerned solely with tree planting. Nurseries were established in the North Island (Whakarewarewa) and in the South Island (Tapanui, Ranfurly, and Hanmer). By 1912 nine associated plantations had been established, three in the North Island and six in the comparatively less forested South Island totalling 27,760 acres [11,239ha]. Whakarewarewa at 9,024 acres [3,653ha] was the largest and the smallest was Dumgeee in Marlborough at 881 acres [357ha].[6] The Forestry Branch trial planted a wide range of European, North American, and Australian tree species, as well as some of the more important indigenous local species, such as Totara (*Podocarpus totara*).[7] They also experimented with dif-

6 *AJHR* C13 (1912): xxiv.

7 The hard won experience of the Forestry Branch about methods of planting and suitability of species was consolidated into a single volume. See Henry Matthews, *Tree-culture in New Zealand* (Wellington: Government Printer, 1905). The efforts of the branch in the South Island were written up for publication overseas; see R. G. Robinson, "Forest-Tree Grow-

ferent methods of planting, including the German continental 'pure-woods system' which mimicked nature whereby trees were planted very closely so that the lower branches died off and knot free timber was produced without any need for pruning. After 25 to 30 years thinning would take place and a decade later rows would be removed to leave 'the best and straightest trees' by now more or less equidistant to each other.[8] On maturation the trees were to be clear felled in strips which would be followed by replanting. From the first, however, some departures from the 'Continental system' of forestry were envisaged; projected demand for railway sleepers was deemed to be sufficiently high to justify planting Tyrolese and Larch Pine at lower densities and pruning them until they were of a suitable size.[9] In 1914, an impressive number of 47 forest species were growing in Forest Branch plantations in the North Island.[10]

Tree planting was labor and time intensive. Seeds had to be collected and grown in nurseries then prepared for transplanting and the actual process of plantation establishment was quite involved. Initially 'pit planting' was favoured whereby the seedlings were transplanted into a pre-prepared trench. Local Maori women worked in the nursery at Whakarewarewa and Maori men and women were employed seasonally in planting out the seedlings. Prison labor was used to help establish the plantations at Waiotapu, Hanmer, and Dumgree. Although prisoners were a 'free' source of labour there were difficulties in achieving the necessary interdepartmental coordination while some of the prisoners worked more effectively than others. The use of prison labor finally ended in 1921.[11]

The official view in 1903 was that the plantations 'cannot be expected to yield mature timber in less than sixty and possibly eighty years'.[12] Unsurprisingly the efforts of the Branch were insufficient to allay concerns about the timber famine. A decade of concerns had magnified to become fears and led to the establishment of a Royal Commission on Forestry in 1913. This was chaired by Henry Haszard, Commissioner of Crown Lands for Westland, a heavily forested region, and its other members comprised builder Sam-

ing in the South Island of New Zealand," *Transactions of the Royal Scottish Arboricultural Society* 38 (1915): 165–178.

8 *AJHR* C 1 (1901): 138.

9 *AJHR* C1 (1901): 138.

10 *AJHR* C1B (1914): 9.

11 *AJHR* C3 (1921): 8.

12 *AJHR* C1 (1903): 8.

uel Clarke, timber manufacturer Charles Murdoch, Frank Lethbridge and Thomas Adams, both farmers (the latter a leading amateur tree planter), and Dr Leonard Cockayne the eminent botanist. There were no professionally trained foresters employed as such in the country at the time. Having toured the country, interviewed witnesses and received submissions, the Commission reported on 10 questions, four of which dealt with indigenous protection and production forests and the remaining six of which concerned exotic plantation forests. These were with regard to afforestation, the probable future demand for timber in New Zealand, the nature and types of timber required, the extent to which the existing state afforestation programme would meet probable demand, the extent to which it ought to be expanded and supplemented, the adequacy of management and control of state plantations and finally under what conditions might the state assist private tree planting.

The Commission considered that existing indigenous timber supplies would last 30 years (i.e. exhaustion by 1943), a slight reduction on the 1909 Forestry Branch projection of 35–40 years,[13] but considerably lower than the estimates of 70 years or 1975 advanced by the Lands Department in 1905.[14] The Commissioners also believed that the 18,870 acres [7,640ha] of existing Forest Branch plantations contained sufficient timber for only 2.6 years. Accordingly they recommended that state planting increase by two-and-a-half times to 6,415 acres per annum [2,596ha]. They were mindful of the need to fully cost afforestation work and drew on *Schlich's Manual of Forestry* in reaching the figure of £7/17/6d per acre for establishing plantations in New Zealand.[15] They also identified eight locations where state plantations ought to be established, displaying a preference for the treeless regions of the South Island, including Central Otago, the McKenzie Basin, the old braided riverbeds of the major Canterbury Rivers, as well as Hanmer, and the Culverden Basin. In the North Island they listed the sand country at the mouth of the Rangitikei River, the Volcanic Plateau to the northeast and southeast of Lake Taupo and the gum lands of the Bay of Islands and Whangaroa.

In reviewing the efforts of the Forestry Branch, the Commissioners were of the view that some trees planted in sizable numbers were clearly unsuccessful. Of this group *Catalpa speciosa* (2,196,544 planted) was 'the worst of all'.[16] Other failures included Larch (10,989,835), the indigenous Totara

13 *AJHR* C4 (1909): 3, 11.
14 *AJHR* C6 (1905): 2.
15 *AJHR* C12 (1913): xxix.
16 *AJHR* C12 (1913): xxxv.

(546,500), European Birch, and Norway Spruce (in total 1,242,723), English Oak (2,041,621), Sycamore (525,247) and Alder (77,912). The Commissioners provided an alternative list of trees for planting, one that varied substantially from the Forestry Branch efforts and included *Pinus radiata*, of which they had planted only 110,161 to 1909,[17] Corscian pine (*Pinus larcicio*), Ponderosa pine, Douglas fir, some eucalypts and various poplars. Last but not least the Commissioners also recommended that a separate Forestry Department be established.

Figure 1: Pinus radiata *was endemic to the Monterey area of California and when introduced to New Zealand, on account of its rapid growth rate, was initially planted for shelterbelts on the open areas of Canterbury and Otago. At first regarded as only suitable for packing and casing, by the early 20th century timber produced from mature small plantations in Canterbury was used for construction purposes. Many of the early plantations were poorly tended and produced overly knotty timber as shown above. It became never-the-less the main species to be planted by both the State Forest Service and the afforestation companies during the 1920s and 1930s. Source: Ebenezer Maxwell, Afforestation in Southern Lands, (Auckland: Withcombe & Tombs, 1930) facing page 64.*

17 *AJHR* C12 (1913): xxxv.

Pinus radiata (formerly described as *Pinus insignis*) sometimes known as Monterey Pine, was a native of California. It was planted in New Zealand as early as 1859 though the exact details of its transmission directly or indirectly from California is unknown (Figure 1). By the 1870s, seeds were received from the UK, Australia, and California, and the tree was initially grown to provide shelter belts on the Canterbury Plains.[18]

While the Commissions' recommendations were not regarded as controversial in New Zealand, the reaction from afar of British colonial foresters ranged from careful criticism to outspoken rejection. Sir William Schlich, the doyen of British Imperial forestry,[19] while acknowledging he had never visited New Zealand, observed that,

> what strikes the reader of the various reports at once is the fact that natural forests have practically been thrown overboard, and that future supplies are to be provided from plantations of exotic trees. This is certainly a very bold measure, which the authorities seem to have adopted because they believe that growth of the indigenous trees is too slow in comparison with that of certain exotic species.[20]

Schlich rightly queried the growth rate data for indigenous and exotic trees. He also raised the question of whether this scale of exotic planting risked 'the development of disease which may lead in the end to disastrous results?'.[21] This point he returned to later in observing that 'exotic trees as a rule [are] more exposed to disease than indigenous trees, and it is impossible to say what diseases the former may develop in the course of time'.[22]

David Hutchins, a Nancy trained forester with experience in Africa and Australia also wrote critically of the Royal Commission on Forestry's report.[23] Hutchins, who later came to New Zealand endeavoured, with some success, to make the case for sustained yield management of indigenous forests and for an independent and professionally staffed forests department, but was

---

18 Winsome Shepherd, "Early Importations of Pinus radiata to New Zealand Part I," *Horticulture New Zealand* 1 (1990): 33–38; and Winsome Shepherd, "Early Importations of Pinus radiata to New Zealand Part II," *Horticulture New Zealand* 1 (1990): 33–35.

19 Anon, "Sir William Schlich," *Journal of Forestry* 25 (1927): 5–8.

20 William Schlich, "Forestry in the Dominion of New Zealand," *Quarterly Journal of Forestry* 1 (1918): 23.

21 Ibid., 24.

22 Ibid., 25.

23 David Hutchins, *A discussion of Australian forestry with special references to forestry in Western Australia* (Perth: Government Printer, 1916).

often his own worst enemy so fierce and emphatic was his advocacy of scientific forestry.[24]

The State Forest Service and the 300,000 acre planting Target

World War I delayed events and it was not until 1919 that a Forests Department separate from the Lands Department was set up and Canadian L.M. Ellis appointed as Director of Forests heading the newly formed State Forest Service. Ellis, a Toronto trained forestry graduate under Bernhard Fernow, had previously worked for Canadian Pacific Railways, served in France in the Canadian Forestry Corps, and was employed briefly by the Scottish Board of Agriculture as an Assistant Forester, before taking up his post in New Zealand.[25] Ellis' preliminary report on forest conditions in New Zealand contained eleven major recommendations.[26] In their breadth they indicated his desire to end the fragmentary and uncoordinated control of forests on Crown Lands and consolidate their administrative oversight in one piece of legislation and within a single department.

The afforestation efforts begun by the Forestry Branch were continued by the State Forest Service, but were not immediately regarded as central to their operations. Initially, at least, Ellis argued for an orthodox forestry model based on sustained yield management of natural forests, with exotic plantation forestry playing a more limited role. This was a reversal of the position that had been adopted when the Forestry Branch began planting in 1897.

One of Ellis' first initiatives was a detailed systemic appraisal of the Dominion's forest resources. Known as the National Forest Inventory and completed from 1921 to 1923, this survey of indigenous forests provided a base line against which planning for sustained yield management could commence. It indicated that timber supplies comprised 38,870 million board

24 Michael Roche, "'The best crop the land will ever grow'; W.F. Massey through the lens of environmental history," *Journal of New Zealand Studies* 8 (2009): 110–112.

25 Michael Roche, "Latter day imperial careering: L.M. Ellis—a Canadian forester in Australia and New Zealand, 1920–1941," *ENNZ: Environment and Nature in New Zealand* 4 (2009): 58–77. http://fennerschool-associated.anu.edu.au/environhist/newzealand/newsletter/

26 *AJHR* C3A (1920): 23.

feet [91.7 million cubic metres] of softwoods and 23,187 million board feet [54.7 million cubic metres] of hardwoods.[27] This data reinforced the concern associated with Ellis' 1925 estimate that sawn timber consumption would reach 675 million board feet [15.9 million cubic metres] per annum by 1965 which would mean the virtual exhaustion of the merchantable indigenous forests. By this same time, State Forest Service estimates suggested that 594,520 acres [240,696ha] of *Pinus radiata* planted over 34 years or 807,260 acres [326,826ha] of *Pinus radiata* and other slower growing species such as Douglas fir or Ponderosa Pine established over 44 years would be needed to offset a future timber famine.[28] Concurrently they instituted a research programme aimed at understanding more about the regeneration and growth rates of the main indigenous timber species; 'searching for the key to Nature's workshop' was how Ellis explained this to his Minister.[29] Although research on Kauri regeneration were encouraging, the situation for the main Podocarp species was less heartening with limited and slow regeneration and potential rotations in the order of 250 to 500 years.

In 1925 as part of a five year review of forest operations, Ellis announced a new plan for 300,000 acres [121,457ha] to be planted in ten years.[30] This was a bold move, though a calculated one, that took him well away from orthodox forest tenets of sustained yield management of natural forest to exotic plantation forestry on an unprecedented scale. Initially, Ellis took the view that the exotic forests would provide a national wood supply and enable the foresters to buy time to solve the problems of indigenous regeneration. In 1925 nine afforestation projects were announced totalling 185,000 acres [74,899ha]. These plantations were to be spread across the country, some such as Riverhead (15,000 acres [6,073ha]) were adjacent to the major urban area of Auckland and all were intended to meet future regional domestic timber needs.[31] From the first, however, the State Forest Service's efforts were concentrated on the Kaingaroa Plains, close to the original Forestry Branch plantings at Waiotapu and one of the areas also identified by the Royal Commission on Forestry.

27 *New Zealand Official Year Book* (Wellington: Government Printer, 1925), 446.

28 Michael Roche, *History of Forestry in New Zealand* (Wellington: Government Print, 1990), 191.

29 Ellis to Bell, September 7, 1921, F6/1/13/1 (Archives New Zealand, Wellington).

30 *AJHR* C3 (1925): 7.

31 L. McIntosh Ellis, "Afforestation," *Auckland Chamber of Commerce Journal* 1 (1925): 8–9.

The major planting effort was increasingly concentrated on the plains of the Volcanic Plateau. An unrecognised cobalt deficiency meant that stock had suffered from 'bush sickness' and progressively lost condition which rendered this large area unsuitable for land settlement. This was particularly important in that when the State Forest Service was increasing and concentrating its planting effort, there was land available which in other circumstances would have been given over to pastoral farming. This colonial settlement ethos otherwise carried over well into the 20th century. It meant that land of even limited potential for farming was given over to that use and that indigenous forest was retained in the long term for future felling only on the poorest of land. Cut over indigenous forest land or other lands unsuited for farming might be considered for plantation forestry. In the longer term plantation forestry tended to be confined to poorer land even after forest economists in the 1960s were able to make a case for the comparative profitability of large scale forestry versus large scale agriculture.[32]

In popular memory the State Forest Service planted only *Pinus radiata,* but this is incorrect. Although the range of species was more restricted than that of the Forests Branch, in part capitalising on the lessons from their efforts, it included a number of Pinus and other species. In 1929 *Pinus radiata, Pinus muricata, Pinus Ponderosa and Pinus laricio* were planted at Kaingaroa, along with Douglas fir and some cupressus species.[33] Kaingaroa was already 29,203 acres [11,823ha] in 1925 and comprising 46 percent of the state plantation forests by area, by 1930 it was 145,963 acres [59094ha] (or 57 percent of the total area planted by the service) far in excess of the initial 80,000 acre target. There was also another 93,280 acres [37,765ha] of land suitable for planting at Kaingaroa at this time. By 1935 it had increased to 242,600 acres [98,174 ha] and amounted to 60 percent of the total area planted by the State Forest Service.[34]

Pinus radiata was the main stay of the planting for a number of reasons. The seed was readily available locally; it could be propagated readily in nurseries, was simply transplanted, grew rapidly and produced a versatile timber. Also critical was the fact that the State Forest Service had been able to reduce the cost of planting from nearly £8 an acre to £1/13/7d per acre by 1924 by using less time-labor-intensive nursery and planting-out techniques.[35]

32 J. T. Ward and E. D. Parkes, *An economic analysis of large-scale land development for agriculture and forestry* (Canterbury: Lincoln College, 1966).

33 *AJHR* C3 (1929): 28.

34 *AJHR* C3 (1925): 17; C3 (1930): 5; C3 (1935): 7.

35 *AJHR* C3 (1924): 9.

Amongst the improvements was the decision taken in 1923, in the light of actual growth rates attained, to replace the 4 x 4 [1.2m x 1.2m] and later 6 x 6 feet [1.8 x 1.8m] planting distance based on European models, with a wider 8 x 8 foot [2.4m x 2.4m] spacing. The State Forest Service also opportunistically was able to secure other Crown Lands which had proven unsuitable for land settlement, such as Gwavas Estate in Hawkes Bay. What became Golden Downs forest in Nelson, and the largest plantation forest in the South Island was established on land where World War I soldier settlement had failed.[36]

Another persistent myth surrounding the state plantation forests is that they were entirely the result of 1930s depression work creation schemes. The impact of the Great Depression had led to the establishment of an Unemployment Board in 1930 to administer relief work and its 'Scheme No. 5' included work on afforestation projects for local councils and state departments. In December 1931 some 800 men were so employed predominantly by the State Forest Service and this rose to over a 1,000 in 1932 and remained in excess of 800 in mid-1935. To this, equivalent numbers were employed part time from mid-1933 to mid-1935. Another 900 were employed under 'Scheme No. 6'. The result of this additional labour power was an extra 100,000 acres [40,486ha] of plantation forest, so that Ellis' original target was reached five years early in 1931 with 307,003 acres [124,293ha] planted.[37] But such an interpretation overlooks Ellis' plans announced in 1925 and the accelerated planting that occurred from this time.

Company Planting 1923 to 1934

Slightly prior to Ellis' announcement of a 300,000 acre [121,457ha] planting target, some in the business community recognised that afforestation could be a lucrative investment. This appreciation was based on the knowledge that *Pinus radiata* could grow to maturity in 25–30 years in New Zealand and that it was possible to promote tree planting as a form of personal retirement income or as an endowment for one's grandchildren. That plantation forest land was taxed at only its unimproved values as if it was grazing

36 Fred Allsop, *The First Fifty Years of New Zealand's Forest Service* (Wellington: Government Printer, 1973), 27.
37 *AJHR* C3 (1931): 4.

land was an added incentive. In 1923 an Auckland land agent/stock broker Landon Smith and his accountant business partner Douglas Wylie registered a small joint stock company called Afforestation Limited in order to plant a 2,800 acre [1,134ha] block of land at Putaruru. Initially the two men had intended to develop the property as farm land, but discussions with H.A. Goudie, formerly the Superintending Nurseryman in the Forests Branch and at the time Conservator of Forests for Rotorua, persuaded them that the land was unsuitable for farming, but would offer much better financial returns if planted in trees.[38]

Figure 2: New Zealand Perpetual Forests advertised heavily in Australia and New Zealand in an effort to encourage people to buy 'bonds'. The company promoted the idea that a £25 bond that would go toward planting an acre of trees which it was confidently predicted would be worth £500 in 25 years when the forest was harvested. This particular advertisement positioned plantation forestry alongside modern scientific farming, drew on 'timber famine' in a muted form of limitless future markets for softwoods, and asserted without any doubt that future financial returns would be considerable. Source: Auckland Chamber of Commerce Journal, 1930. Author's collection.

38 Brian Healy, *A Hundred Million Pine Trees* (Auckland: Hodder and Stoughton, 1982), 11–12.

Heartened by the ease of selling shares in Afforestation Limited, Smith and Wylie set up New Zealand Perpetual Forests and Putaruru Forests Limited later in that same year. These were much larger companies, but instead of conventionally selling shares in a joint stock company to investors, Smith and Wylie made use of a new means of raising capital whereby individuals bought what was commonly and confusingly termed a 'bond' in return for which the company would plant and tend an acre of forest (680 trees to the acre at 8 x 8 foot [1.8 x 1.8m] spacing) and on maturity harvest them and return funds to the bond holder. The typical promissory formula was a £25 bond for £500 in 25 years (Figure 2). Bond companies had three main features including a private parent company typically in the hands of a few shareholders which made its profits from the actual task of planting the forests, a body of bond holders with no control over the parent company but with some rights over the forests once planted, and trustees appointed to look after the bond holders' interests. Other afforestation companies were quickly registered in order to mimic New Zealand Perpetual Forests; 30 were set up from 1923 to 1925, mainly Auckland based, they peaked in number at around 40 by 1933.

Door-to-door bond selling meant that capital was raised from across the country for plantations that were principally located in the central North Island. Some companies saw an opportunity to raise still further funds offshore. New Zealand Perpetual Forests, for instance, ran large advertising campaigns in Australia. Others were more ambitious; Commercial Pine Forests Ltd raised 95 percent of its bond capital from India and Sri Lanka. New Zealand Redwood Forests also targeted India. Australian businessmen even established Afforestation Propriety Limited in order to tap into the bond market and plant land in New Zealand. Timberlands New Zealand Limited was also another Australian registered company planting solely in New Zealand. An authoritative estimate was that 60 percent or £5 million of bond money was raised from outside New Zealand.[39] The claimed growth rates and likely financial returns caused the State Forest Service considerable anxiety, so there was an uneasy relationship between them and the afforestation companies.

39 J. Melville, "Co-operative afforestation and the protection of forestry investors," *Timber Growers Quarterly Review* 2 (1930): 23.

New Zealand Perpetual Forests

New Zealand Perpetual Forests was the largest and most active of the afforestation companies.[40] In 1925 they recruited Owen Jones, a professionally trained forester to oversee their planting operations. Jones was in many ways an archetypical British colonial forester having completed a degree in natural sciences in 1910 and a Diploma of Forestry in 1911 at Oxford University and then served as an Assistant Forester in Ceylon (Sri Lanka). After war service he went to Australia where he headed the Victorian Forestry Commission from 1919 to 1925. In 1923 he led the Australian delegation to the Empire Forestry Conference in Canada. Dissatisfied with public service forestry he accepted a position with New Zealand Perpetual Forests as their 'Forestry Administrator' remaining with the company until 1940.

Jones described the activities of New Zealand Perpetual Forests in the *Empire Forestry Journal* of 1928. He began by outlining the character of the pumice lands of the volcanic plateau, at an altitude of 600m plus, covered in tussock grass *(Poa cita)* and low manuka (*Leptospermum scoparium*) scrub. The land, he noted,

> for settlement purposes it is of small value, much of it being afflicted with mysterious "bush sickness" which often proves fatal to cattle or sheep that are kept upon it for more than a few months at a time, while most of the attempts to crop it have ended in failure. It has however, proved suitable for the growth of softwood species, and there is now every prospect that this vast area, hitherto abandoned almost entirely to the wild horse and the wild pig will blossom under the hand of the forester into an important tree-producing regions.[41]

Jones then outlined the establishment of a field headquarters and nursery, the survey and subdivision of land into compartments of 250–300 acres [101–121ha] separated for planting. Some 70 miles [113 km] of road were

40 'Perpetual' was the name deliberately chosen by the company founders in 1923 to signal their intention that the forests as they successively came to maturity, would be felled and that regeneration would take place naturally or assisted by further planting. See Healy, *A Hundred Million Trees*, 21. It may not be a coincidence that Ellis in speaking about the need for state forestry more generally in New Zealand in 1921 and 1922 was reported in the press on several occasions as using the phrase 'perpetuation of the use' of New Zealand's forests. For example see "Forest Conservation," *Auckland Star*, October 31, 1921; "Timber Hunger," *Evening Post*, November 16, 1921; "A Forestry Boom," *Northern Advocate*, January 6, 1922; "Forestry at work," *Evening Post*, July 24, 1922.

41 Owen Jones, "An Afforestation Company's Operations in New Zealand," *Empire Forestry Journal* 7 (1928): 66.

formed in the 43,000 acre [17,409ha] Maraetai Block and the plantings were separated by regular firebreaks. The 600–700 seasonal labourers were housed in a number of specially built camps. The pay according to Jones was 2/- above the standard wage.

Jones noted that *Pinus radiata* was the principal species planted by the company 'by reason of its remarkably rapid growth, its hardiness, its adaptability, and the general all round utility of its timber'.[42] He continued, however, echoing his professional training that, 'it is realised however, that it is not desirable to have excessively large areas of one species only, and also that the best results can be achieved by establishing in each locality the species best adapted to it'.[43] This meant that the company planted some land in the more frost resistant Ponderosa pine, along with small areas in Douglas fir, Redwoods, Weymouth pine, Maritime pine, Corsican pine, Monterey cypress and some poplar species. The length of the list tends to obscure the comparative concentration, New Zealand Redwood Forests excepted, of the companies on planting *Pinus radiata.* By 1927 New Zealand Perpetual Forests had 29.7 million trees of which 70 percent were *Pinus radiata*, 18 percent *Pinus ponderosa,* 7 percent Douglas fir, and the remainder in *Sequoia sempervirens*, Macrocarpa, *pinus strobus* and *Pinis pinaster*[44]. Over time the percentage of *Pinus radiata* rose to be in excess of 80 percent. Three of these species; *Pinus radiata,* Ponderosa pine, and Douglas fir had been recommended by the Royal Commission on Forestry. Driven by economic imperatives New Zealand Perpetual Forests and the afforestation companies more generally moved quite quickly toward creating a plantation monoculture.

New Zealand Perpetual Forests had also developed improved simplified nursery practices, abandoning labor intensive hand sowing in beds protected by hessian covering in favour of open drilling of seeds. When ready for replanting the best seedlings in batches of 1000 to 1200 were taken in company trucks to depots and heeled in to await collection by planters. At the height of the extended planting season of five to six months over 300,000 trees were required daily.[45] In 1927 New Zealand Perpetual Forests shifted from 8

42 Jones, "An Afforestation Company's Operations in New Zealand," 70.

43 Ibid., 71.

44 H. Hugh Corbin, "Progress Report No 5 on the Forestry Operation of New Zealand Perpetual Forests Limited, 1927," (copy in F1 423 29/5/4 part 2 Archives New Zealand, Wellington).

45 Jones, "An Afforestation Company's Operations," 72.

x 8 [1.8 x 1.8m] to 9 x 9 [2.7 x 2.7m] planting in view of the *Pinus radiata* growth rates (3 to 3.7m in 3 years).

Paternalistically Jones also commented of the workforce that,

> A considerable proportion of the labour force, particularly amongst the actual tree-planters, are Maoris. The Maoris take a real interest in tree-planting and provided they are properly handled and due attention is paid to their little peculiarities, which they have no less than the rest of mankind, they render satisfactory service. Being naturally of a musical disposition they experience little difficulty in amusing themselves during the evenings, no small factor in promoting contentment in camp life; they accept without undue grumblings the discomforts inevitable during cold, wet weather; their mercurial temperament is seldom downcast for long, and they can and do smile cheerfully under most circumstances.[46]

In 1926, New Zealand Perpetual Forests planted 30 million trees on 46,000 acres [18,623ha] of land with a work force of 630. Jones heralded this effort as 'a record one for the British Empire'.[47] Looking ahead he proclaimed that 'In the near future the areas now established will present an impressive spectacle, changing a bleak and dismal countryside into a thing of beauty and converting lands hitherto waste and valueless into an asset of the utmost national importance'.[48] Although Jones was important to the company in implementing its planting programme he remained somewhat on the outer, especially from the late 1930s when the other directors turned their attention more to utilisation including construction of their own sawmill and schemes for a pulp mill. Jones' plan for the careful silvicultural tending of the plantations was rejected, to the extent that he was not re-employed by them in 1945 after war service in the New Zealand forestry company.

The bondholding system underpinning company afforestation was open to abuse. For instance, afforestation company directors purchased lands for planting at inflated prices from other companies in which they had a financial interest. In other cases the cost of planting was excessive and the company providing the seedlings was separately owned by the directors of the main afforestation company.[49] These irregularities led to the appointment of a Commission of Inquiry and new legislation that made bond selling il-

46 Jones, "An Afforestation Company's Operations in New Zealand," 69.

47 Owen Jones, "Forestry Administrator's Report 1926 New Zealand Perpetual Forest Ltd, 1926," n.p.

48 Ibid.

49 Horace Belshaw and Frank Stephens, "The financing of afforestation, flax, tobacco and Tung oil companies," *Economic Record* 8 (1932): 251–254.

legal. The existing bond selling afforestation companies were under separate legislation, forced to dissolve themselves with the larger of them incorporating themselves as ordinary joint stock companies with the former bond holders now being shareholders. Thus New Zealand Perpetual Forests became reorganized as New Zealand Forest Products in 1934. The demise of the bond selling companies virtually ended all additional company planting until the 1960s. Total company planting situated mainly on the volcanic plateau by this time amounted to 272,336 acres [110,257ha] in 1934.[50] New Zealand Perpetual Forests was the single largest owner with some 172,307 acres [69,760ha] planted in 1937.[51]

The State Forest Service planting effort was also wound down at this time. Ellis, won over to the possibilities of plantation forestry in New Zealand had advocated a target of 5,000,000 acres [2,023,390 ha], before he departed in 1928 to take up a position with an Australian afforestation company, but this was never officially adopted.[52] He was also mindful of how difficult it would be to persuade the politicians to plant steadily for a complete cycle of 25 plus years, and thus had advocated a decade long program, aware that the narrow range of age classes would bequeath some forest management problems to his successors. State planting, though accelerated by depression work creation schemes, was wound down in the mid-1930s. By this time Ellis's original 300,000 acre (121,403ha) target had been exceeded; the state planted 393,998 acres (159,442ha) to 1934 and once the company plantings of 272,336 acres (110,613ha) to that same year were also included these were considered to provide sufficient wood supplies for estimated demand in the 1960s.[53] Other developments would in any case have made it difficult to continue with large scale planting, for from 1932 DSIR [Department of Scientific and Industrial Research] research had offered a solution to the problem of 'bush sickness', by identifying the cobalt deficiency in the volcanic soils which meant that lands previously regarded as unsuitable for land settlement now became reincorporated within the agricultural and pastoral frontier and away from forestry.[54] A.D. McGavock, Director of Forests 1932–1938, also sought in 1934 to bring a 'proper focus' to national forest

50 *New Zealand Official Year Book* (Wellington: Government Printer, 1936), 378.

51 Healy, *A Hundred Million Pine Trees*, 85.

52 Roche, "Latter day imperial careering".

53 *AJHR* C3 (1934): 6; and *New Zealand Official Year Book*, 1936, 378.

54 Ross Galbraeth, *DSIR Making Science Work for New Zealand* (Wellington: Victoria University Press, 1998), 29–31.

policy away from the 'false perspective' of exotic afforestation and back to a situation where more attention was paid to protection forestry and the indigenous production forestry.[55] The challenge, as he perceived it, was one of managing the 'over mature' indigenous forests in ways that promoted the highest possible levels of net growth.

Post World War II attention turned to forest utilisation. In 1939 the new Director of Forests, Alex Entrican, an engineer by background, saw his mission as one of developing and implementing a national timber and pulp and paper mill scheme for the plantations created in the 1920s and 1930s. The private sector responded more quickly with Whakatane Board Mills beginning production in 1939 and from the late 1940s New Zealand Forests Products also began to develop utilization plans for its own forests culminating in the opening of its Kinleith pulp and paper plant in 1954. The tender for state grown wood from Kaingaroa was won by Tasman Pulp and Paper and a pulp and paper mill opened at Kawerau in 1954.[56]

Plantation Forestry and Risk

Schlich had pointed to the potential vulnerability of plantation forestry in New Zealand to disease.[57] The Forest Branch had tended to be more concerned about the danger posed by fire and by 1914 had a system of huts and patrols in place at the main Whakarewarewa plantation. That same year an acre [0.47ha] of six-year-old Austrian pines had been damaged but not destroyed in fire that spread from a careless nearby road gang.[58] Other animal pests and disease threats were acknowledged, but fire tended to be the major concern, possibly because it was then routinely used in clearing indigenous forest land for the expansion of farming at the rural fringe while sparks from steam locomotives were a further cause of rural fires.

Ellis brought a heightened concern about fire danger with him from Canada, where he had experienced forest fires first hand when working for Canadian Pacific Railways. He also introduced some Canadian models for

55 *AJHR* C3 (1934): 2.

56 Healy, "A Hundred Million Pine Trees," 132–139; Thomas Simpson, *Kauri to Radiata* (Auckland: Hodder and Stoughton, 1973), 335–340.

57 Schlich, "Forestry in the Dominion of New Zealand".

58 *AJHR* C1b (1914): 8.

fire prevention and adapted other ideas from Western Australia which were incorporated into the new Forests Act of 1921–22. Thus the State Forest Service quickly instituted 'Fire Districts' in which there were restrictions on fire lighting, a requirement for all males over 16 to assist in the event of fire, and honorary rangers on patrol.[59] Initially these were geared to preventing fire in the indigenous forests, which in dry years could burn, but over time the focus began to change towards closer protection of the plantation forests, especially as they contained the future stock of timber and pulpwood. This concern was shared by the afforestation companies who could easily imagine their entire forest estate being destroyed in a single conflagration. Dry weather in 1924 assisted state tree planting in Canterbury but simultaneously caused Ellis visiting the region to ask for 'extraordinary vigilance' and public cooperation to prevent accidental fires and their spread into the forests.[60]

Amending legislation in 1925 extended fire district provisions to encompass both state and company plantations. By this time the State Forest Service had set up a network of fire observation posts, purchased some basic fire fighting equipment and was making use of psychrometers to measure relative humidity in order to better understand 'fire seasons'.[61] It was recognised, however, that wet years significantly reduced the risk of forest fires, which in any case State Forest Service data revealed was more likely to occur in indigenous forests adjacent to settled areas as a result of burning off.[62] Fires caused minimal damage in state plantations until November 1927 when 60 acres [24ha] were destroyed at Hanmer state forest, in 1929 30 acres [12ha] were consumed at Kaingaroa and 1930 another 132 acres [53ha] were lost at Kaingaroa and 252 [102ha] at Longwood state forest respectively.[63] The Hanmer fire late in 1927 had destroyed near two percent of the plantation by area of largely ten-year-old trees. The State Forest Service internal report noted that with gale force winds the fire had easily jumped the 66ft (60m) firebreaks, though these did provide crucial access for fire fighters. Motorised pumping equipment proved effective, though some modifications were sug-

59 Helen Beaglehole, *Fire in the Hills. A history of rural fire-fighting in New Zealand* (Christchurch: Canterbury University Press, 2012).

60 *Northern Advocate*, January 22, 1924.

61 *AJHR* C3 (1926): 15.

62 The quite elaborate State Forest Service statistics from 1921 do not, however, separate indigenous production forest from plantations.

63 *AJHR* C3 (1928): 11; C3 (1929): 10; and C3 (1930): 8.

gested, while foreshadowing later deficiencies maintaining telephone communications proved difficult.[64]

Afforestation companies recorded losses of 1016 acres [411ha] in four separate fires in 1926.[65] New Zealand Forest Products suffered again in 1937 when 564 acres [228ha] of a 700 acre [283ha] block of 11-year-old 50 foot [15m] high *Pinus radiata* plantation near to Upper Atiamuri were destroyed by fire. The fire was caused by a sudden wind change which sent sparks from the burning off of scrub and fern on adjacent farmland into the plantation. Dry conditions meant that it easily took hold. Over 150 men from the company, State Forest Service, the Public Works Department, as well as nearby settlers fought the blaze. Jones successfully used a plane to search for the seat of the fire and direct fire fighting efforts.[66] The company's firebreaks and forest road system proved important in controlling the blaze which had posed a serious threat as it occurred on the edge of a much larger contiguous area of plantations.[67]

The first really sizable plantation forest fire did not take place until 1955 when 7390 acres [2290ha] in excess of 40 percent of the plantation, equating to 4,705,000 cubic feet [133227 cu m] of wood, was burnt at Balmoral State Forest on the Canterbury plains. The fire started in the old mill adjacent to the plantation and spread into a forest that was vulnerable to fire damage because of severe wind damage in 1945 (from which much debris remained on the forest floor), a looper caterpillar infestation (*Selidosema suavis*) in 1952 which killed 750 acres [303ha] of *Pinus radiata*, and heavy thinning of the *Pinus laricio* stands so that it contained much inflammable material. Although the fire was initially contained, high winds and changes of wind direction meant that it was unable to remain so. A Commission of Inquiry was held two months later into how this, the largest fire in an exotic state forest had taken place and with a view to identifying new fire fighting practices. Witnesses during the inquiry pointed to the limitations of telephone communication, runners having had to be resorted to, and pressed for use of radios. The Commission itself identified errors of judgement on the part of local Forest Service staff up to level of the Conservator of Forests in the early response to the fire, though no disciplinary measures were deemed necessary

64 *Evening Post*, November 17, 1927.

65 *AJHR* C3 (1926): 15.

66 *Auckland Star*, December 2, 1937.

67 *Auckland Star*, Sept 10, 1938.

and set out a number of recommendations.[68] A paper subsequently published in the *New Zealand Journal of Forestry* highlighted the vulnerability of such even aged class plantations to fire damage and the ease with which crown fires jumped across fire breaks.[69] Entrican described the Balmoral fires as 'a grievous national loss' and suggested despite the mild tone of the commission of inquiry report that it 'had taught hard lessons' to the New Zealand Forest Service.[70]

The huge fatal plantation forest fire of national significance never occurred. Arguably this was testimony to the early recognition of the threat and the effort put into fire protection. Plantations as a matter of course had firebreaks, and fire protection was an area of rare and unstinting cooperation between the State Forest Service and the afforestation companies.

The State Forest Service respected Schlich's warnings about the risk of insect and fungoid pests in plantation forests. For a time geographic isolation and the time taken for plant material and potentially damaging insects to travel to New Zealand provided a degree of protection. But as early as 1923 the Tortrix moth was causing localised problems and some Eucalypts were also suffering from damage from Australian insects.[71] It was appreciated that improved inspection and quarantine procedures were required to keep insect pests from entering the country. However the State Forest Service was alarmed that same year to discover that it was unable legally to prevent the importing of insect infested logs and other timbers. Accordingly discussions were opened with the Department of Agriculture, which was responsible for other animal quarantine matters, in order to take a coordinated approach and seek to use the *Customs Act* in order to reject insect infested timber. The necessary regulations were drafted in 1924 and subsequently revised in 1925 and 1928 as the matters became mired in interdepartmental exchanges. This was not aided by the NSW government intimating that they would be unable to issue certificates that their export logs and timbers were insect free as the New Zealand government had wished. By this time the proposed regulation had been so modified as to be, in the view of one senior forester,

68 Inquiries/investigation—Balmoral Forest Fire—November 1955—inquiry and minutes of evidence. F1 W3129 353 IN 55.1.1 (Archives New Zealand, Wellington); and Report of the Commission of Inquiry into the Balmoral Forest Fire (Typescript author's collection).

69 Keith W Prior, "The Balmoral Forest fire," *New Zealand Journal of Forestry* 7 (1958): 35–50.

70 *AHJR* C3 (1956): 5. The State Forest Service was renamed the New Zealand Forest Service in 1949.

71 *AJHR* C3 (1923): 16.

of doubtful value. While the State Forest Service presented the plantations as 'at present unprotected from exotic insect attack' all that was effectively achieved was the agreement of the Australians to export only debarked logs.[72] Most of this group of insect pests were, however, specific to Eucalyptus species and fortunately did not pose a threat to the pine plantations. In 1927 insect pests were discovered in hardwood timbers imported from Australia, in Oak, Ash and Redwood timbers from North America and in timbers and slats from Europe. Of the various species, the bark beetle was potentially the most damaging to conifers.[73] The State Forest Service again pressed for regulations to prevent imported insects and other diseases in 1940, but this time under a provision in the Forests Act. This was rejected by the Crown Law Office as too dramatic in its scope for use in a regulation and too invasive of private property rights, thus effective inspection of log imports was delayed until after WWII.

From 1927 to 1928 the State Forest Service carried out a forest insect survey with the assistance of the Department of Agriculture. This revealed the wood wasp *Sirex noctilio* (then described as *juvencus*) as a particularly unwelcome arrival from the UK where it was not a major pest. It had probably appeared as early as 1900.[74] Overall the results of the survey were alarming with the condition of the forests described as 'very unsatisfactory' from an entomological point of view.[75] A Forest Biological station was established at the Cawthorn Institute in 1929 after the afforestation companies association had approached the DSIR to help fund the research. The State Forest Service seconded an entomologist and a fuller national insect survey was commenced, but severe public sector funding cuts as part of depression retrenchment measures slowed progress. Opinions were also somewhat divided at this point with a Canterbury University College entomologist claiming the plantations in Canterbury were 'swarming' with Sirex while the regional Conservator of Forests argued that Sirex only attacked weak or dead trees that would be removed from the plantation in any case. The

72 Proposal for Order-in-Council to control Imported Insects dangerous to New Zealand Forests and Forest Products, n.d. F1 151 1/1/2/1 part 1, Forest Protection (Insect and Disease) Regulations 1923–1950 (Archives New Zealand, Wellington).

73 *AJHR* C3 (1927): 21.

74 A. Clark, "The infestation of sirex juvencus in Canterbury," *Te Kura Ngahere* 2 (1927): 10–16; George Rawlings, "Recent observations on the sirex noctilo population in Pinus radiata forest in New Zealand," *New Zealand Journal of Forestry* 5 (1948): 411–421.

75 *AJHR* C3 (1929): 14.

borer was conceded 'prevalent' but did not constitute a menace.[76] Behind the official position, concerns were somewhat greater than publically acknowledged. In keeping with the approach of the time, biological control was the favored response.[77] A parasite was imported from the UK with the assistance of the Imperial Institute of Entomological in 1932 and distributed throughout company and state plantations. Jones at New Zealand Perpetual Forests received a dozen of these female *Rhyssa persusoria* in 1933 for release in company plantations. The State Forest Service released more of the parasites in Canterbury plantations and at Whakarewarewa in 1935.[78] With the establishment of the Forest Research Institute in 1947 more systematic entomological research commenced.[79]

The more serious Sirex outbreak did not occur until 1948 when it was observed in the major plantation forests at Waiotapu and Whakarewarewa. By this time entomological research had identified that fungal spores on the Sirex eggs were having a greater impact on the trees than Sirex itself and that the three year life cycle of the wasp could produce a sudden build-up of numbers. *Ibalia leucospoides* was now also considered to be a more effective parasite in the prevailing conditions than *Rhyssa.*[80] Forest Research Institute pathologist George Rawlings ruled out chemical control on the grounds of cost, believed that biological control ought to be investigated, but suggested that silvicultural control (salvaging dead logs and carefully timed thinnings) held the best long term hopes for a solution. Reviewing the question of epidemics in *Pinus radiata* forests in the mid-1950s he offered the view that it was not the epidemics that were the real problem, for 'they are spectacular, but they are transitory in nature … of greater significance is the danger of introducing a new species which will be capable of maintaining its population at an epidemic level until starved out by lack of food, or if it is an insect, until

76 *Christchurch Press*, June 14, 1932.

77 There was however a contest within DSIR and Cawthron, a privately endowed research institute, over biological control versus chemical treatments. In the case of the fruit growing sector the chemical spraying advocates triumphed over the supporters of biological control. See Michael Roche, "'Wilderness to Orchard': The Export Apple Industry in Nelson, New Zealand 1908–1940," *Environment and History* 9 (2003): 435–450.

78 F 1 575 50/2/4 part 1, Sirex Noctilio—General 1931–1949 (Archives New Zealand, Wellington).

79 John A. Kininmonth, *A history of forestry research in New Zealand* (Rotorua: New Zealand Forest Research Institute, 1997), 95.

80 F 1 575 50/2/4 part 1, Sirex Noctilio—General 1931–1949 (Archives New Zealand, Wellington).

the balance is restored through the introduction of parasites'.[81] Entrican, the Director of Forests, was even more sanguine, declaring in 1955 that 'it was entirely a matter of luck' that the sirex epidemic and another small one of indigenous Looper caterpillar (*Pseudocoremia suavis*) in another plantation (Eyrewell State Forest) in 1951–52, 'did not result in catastrophic damage'.[82]

In the aftermath of the Sirex outbreak and the Looper caterpillar infestation Entrican invited the well-known Canadian forest pathologist J.J. De Gryse to report on the New Zealand situation in 1953. De Gryse's report displayed confidence in biological control to combat insect epidemics but it also confirmed that Sirex outbreaks were a direct legacy of the conditions under which the plantations were established in the 1920s. The decade long planting boom had produced forests of relatively even age classes and the lack of thinning and other silvicultural treatment in subsequent decades produced an environment ideal for Sirex numbers to explode, especially after successive dry years and when parts of the forest had suffered wind damage.[83] In the longer term, as Rawlings suggested, more intensive silvicultural management of planation forests effectively quelled the Sirex threat. The forestry profession endorsed De Gryse's report and took as its four messages the need for an effective quarantine system (heightened by the expansion of commercial air services), on-going monitoring of forest insect and fungi populations, fundamental research in forest biology, and higher standards of silviculture.[84]

A second forestry planting effort from 1960 to the early 1980s and a wider range of age classes and more intensive silvicultural treatment of plantations for saw log production reduced the fire and insect risk. Environmentalists opposed to the on-going harvesting of indigenous forests did, however, now express some concerns about the vulnerability of plantation forestry monocultures. In response Forest Research Scientists claimed that while monocultures contained an element of risk it was not necessarily higher than in a natural forest, and that planting species other than *Pinus radiata* would only spread the risk but not reduce it.[85]

81 George Rawlings, "Epidemics in Pinus radiata forests in New Zealand," *New Zealand Journal of Forestry* 7 (1955): 53–55.

82 Alex Entrican, Foreword in Joseph J. De Gryse, *Forest Pathology in New Zealand* (Wellington, New Zealand Forest Service, 1955), Bulletin No. 11. [cover misprinted as Bulletin No. 1].

83 Joseph J. De Gryse, *Forest Pathology in New Zealand*, 4.

84 Frank Hutchinson, "Presidential Review," *New Zealand Journal of Forestry* 7 (1956): 12–16.

85 J. Bain, "Forest monocultures—how safe are they?," *New Zealand Journal of Forestry* 26 (1981): 37–42; C.K. Chou, "Monoculture, species diversification and disease hazards in

Conclusion

Butlin signaled as key features of the 19th century plantation agriculture that they are large establishments, with significant capital requirements, produced specialised crops for export, and that they tended to be controlled by metropolitan interests. Plantation forestry in New Zealand has displayed some of these attributes. The colonial state had suitable land available, especially in the cobalt deficient lands of the central North Island, so that this area was not sought after for agriculture which otherwise tended to have priority as a land use. Kaingaroa eventually became the largest contiguous forest plantation in the Southern Hemisphere.[86] 'Free labor' was also available to the state when it was using prison inmates to plant and something of a reserve labor force existed in the form of rural Maori in the Rotorua area while the work schemes of the Great Depression enabled much more to be planted than originally planned. The main point of departure from Butlin is that the controlling interests were largely New Zealand based in that over half of the plantations were established by the state in the 1920s and 1930s[87]. Even when the company planting is considered, this was largely New Zealand domiciled, although there was some Australian involvement. The bond selling format actually gave New Zealand entrepreneurs the opportunity to draw in capital from outside New Zealand to fund their schemes. The plantation forests were also originally intended for local consumption, but the scale of planting supported forestry exports in the form of paper and later logs in the 1950s and 1960s.

Also of interest in the New Zealand setting is the fashion in which professional foresters seized on the possibilities of large scale plantation forestry in a manner that was beyond the bounds of their European and North American training. One consequence of the new nursery and planting techniques was a greater concentration on one species, the hitherto obscure *Pinus radiata,* particularly by companies, but also by the state after 1925, so that plantation forestry increasingly took the form of an exotic monoculture. The narrow age class range and lack of silvicultural treatment, particularly thinning, in the plantations established in the 1920s and early 1930s made them vulnerable to both fire and insect attack. Fire risk was readily reduced by in-

forestry," *New Zealand Journal of Forestry* 26 (1981): 20–36.

86 A. Priestly Thomson, "Why sell Kaingaroa Forest?," *N.Z. Forestry* 40 (1997): 4.

87 Though after the sale of cutting rights to the state forests in the late 1980s and early 1990s overseas interests have come to dominate the ownership of the plantation forest estate.

corporation of fire breaks during planting, the purchase of specialized equipment, imposition of fire bans and putting in place of a legislative regime. Dry seasons and unexpected wind shifts provided the macro and micro- environmental conditions that produced some of the more damaging plantation forest fires. The plantations were also vulnerable to insect attack which was initially countered by efforts at biological control, though in the longer term more intensive silvicultural treatment in the from of thinning and pruning and the removal of any dead and dying trees along with some aerial spraying reduced the threat. Sirex, endemic to the UK where it was a minor pest, caused the greatest threat to *Pinus radiata*, the American species that dominated the New Zealand plantation forestry sector. Looper caterpillar was the only indigenous insect to pose a significant threat to the plantations. It was acknowledged at the time, at the highest levels of the New Zealand Forest Service, that to some extent nothing more than good fortune had prevented the Sirex from becoming a national disaster. Strenuous border biosecurity efforts since that time have meant that worst case scenarios have not been realized in New Zealand.

While there are distinctive elements to the New Zealand plantation forest experience it is not a singular example. Australia had at approximately the same time followed its own trajectory in creating a sizable softwood plantation forest estate. There are also large areas of *Pinus radiata* and other soft wood plantations in South Africa and Chile. A comparative environmental history of all of these southern hemisphere softwood plantations has merit. There are now plantation forests in various tropical countries, arguably these have more connections with some of the plantation system characteristics outlined by Butlin? New Zealand's era of plantation forest establishment in the 1920s can be linked to timber famine concerns on the part of the state and the opportunity for financial returns on the part of the companies. It capitalized on the presence of *Pinus radiata* brought from California, and on the boldness and calculated risk taking of Ellis as Director of Forests in the 1920s and the professional skills of company foresters such as Jones.

How Nature Works: Business, Ecology, and Rubber Plantations in Colonial Southeast Asia, 1919–1939

Michitake Aso

In 1937, Jean Adam, a Paris-based expert, published a twelve-page article calling for a "colonial ecology." Adam began by defining ecology as the study of "the relations existing between living organism and the surrounding environment." Next, he described the goals of the sub-discipline of agricultural ecology: "[it] has as its purpose the adaptation of the use of plants and of animals to the agricultural environment, with the goal of obtaining, under the best conditions, the highest possible yield of products helpful to mankind." While agricultural societies could benefit from such knowledge, Adam argued that by focusing too narrowly on non-human nature, the "physical factors, that constitute the conditions of material life," agricultural ecology had limited its scope of action.[1]

In order to redress the imbalance between studies of natural and human environments, and the preponderance of what he termed "technical" over "moral" elites, Adam sketched out his vision of a colonial ecology. He argued that this "science" could offer solutions to the interrelated problems of both the natural and human worlds in the colonies. Such an ecology would "study the colonial environment [*milieu*], natural and human,—study the tools of production, plant and animal, to which it would be good to add the underground resources, that equally have something to contribute,—study man, European and Indigenous, in his action in the colonial environment." Through this research, colonial ecology would serve as "the summary freeing the mastered ideas from the chapters where the details are given, to scan the horizon and grasp in a synthetic view the generality of the ideas to be acquired."

Published under the Popular Front government in France, which concerned itself with the conditions of the colonized, Adam's article emphasized "the well-being [*bien-être*] of populations." Adam noted that as a colonial

1 Jean Adam, "De l'écologie agricole à l'écologie coloniale," *L'Agronomie Coloniale* 237 (1937): 2, 3, 11.

power, France was responsible for the "betterment [*mieux-être*], material and moral" of the indigenous peoples. Within the imperial project, colonial ecology could "show the indispensible connections and serve as a preface for courses teaching the sciences of colonial specializations." This science had the added benefit that it could then be reincorporated into agricultural practices in France. Yet, even as Paris-based Adam sought to incorporate human culture into ecology, in colonial Southeast Asia this science remained oriented towards environmental management and business needs.[2]

Throughout the colonial world of the early twentieth century, experts such as agronomists, engineers, and medical doctors developed and applied sciences to increase and improve yields of plants and animals for human consumption. Planters and herders drew from economic botany, plant pathology, and veterinary medicine in order better to understand and manipulate plant and animal natures.[3] Less scholarly attention has been paid to the role of ecology in asserting control over nature in European empires. Through a study of the rubber industry in French Indochina, now Vietnam, Cambodia, and Laos, and British Malaya, now Malaysia, this article investigates the role of plantation agriculture in encouraging the growth of networks of people and institutions that produced and disseminated ecological knowledge. A history of the rubber industry demonstrates that the theoretical aspirations of ecology were not antithetical to the practice-oriented ways of plantation agriculture.

Historians of ecology have illuminated the role that universities, natural history museums, and national parks played in the development of this discipline in Europe and the United States.[4] In the colonial world, scholars have

2 Ibid.: 11.

3 See for example John Soluri, *Banana Cultures: Agriculture, Consumption, and Environmental Change in Honduras and the United States* (Austin: University of Texas Press, 2005); James McCann, *Maize and Grace: History, Corn and Africa's New Landscapes, 1500–1999* (Boston, MA: African Studies Center, Boston University, 1999); Christophe Bonneuil, *Mettre en ordre et discipliner les tropiques : les sciences du végétal dans l'empire français, 1870–1940* (Paris: Université Paris VII–Denis Diderot, 1997); Karen Brown and Daniel Gilfoyle, *Healing the Herds: Disease, Livestock Economies, and the Globalization of Veterinary Medicine* (Athens: Ohio University Press, 2010).

4 Donald Worster, *Nature's Economy: A History of Ecological Ideas* (New York: Cambridge University Press, 1994); Gregg Mitman, *The State of Nature: Ecology, Community, and American Social Thought, 1900–1950* (Chicago: University of Chicago Press, 1992); Sharon E. Kingsland, *The Evolution of American Ecology, 1890–2000* (Baltimore: Johns Hopkins University Press, 2005); Eugene Cittadino, *Nature as the Laboratory: Darwinian Plant Ecology in the German Empire, 1880–1900* (New York: Cambridge University Press, 1990);

identified botanic gardens and forests as places of ecological knowledge.[5] This article seeks to contribute to a global history of ecology by showing how plantation agriculture in general and rubber plantations in particular both aided in the codification of ecology in colonial Southeast Asia in the 1920s and provided one of the key motivations to do so. Initially, colonial experts and their associated institutions, such as experimental stations and hospitals, served as producers and transmitters of ecology. These experts helped to introduce ecological concepts as they moved through scientific, government, and business circles. As many colonial officials and planters increasingly found ecology to be a useful language with which to communicate, these actors attempted to frame ecology as an "apolitical" science in the highly politicized space of the colonial empires.[6]

Because of these efforts to depoliticize ecology, ecological knowledge produced in French Indochina and British Malaya in the 1920s and 1930s generally took little account of "human environments" and focused instead on biophysical environments, or the interactions among plants, animals, soils, climates, and waters. There were isolated cases of interest in "native" agriculture, but these were exceptions that proved the rule. In other words, unlike British colonial Africa of the 1930s, where Helen Tilley has shown how ecological sciences and social anthropology cross-fertilized to generate new understandings of human and non-human environments, ecology in much of Southeast Asia continued to depend on a model of generic and rational farmers.[7] Only near the end of the colonial period did a few human geographers, agricultural agents, and engineers, attempt to incorporate cultural understandings of agricultural practices into the production of scientific knowledge.

Lynn K. Nyhart, *Modern Nature: The Rise of the Biological Perspective in Germany* (Chicago: University of Chicago Press, 2009).

5 Andrew Goss, *The Floracrats: State-Sponsored Science and the Failure of the Enlightenment in Indonesia* (Madison, WI: University of Wisconsin Press, 2011); Frederic Thomas, "Ecologie et gestion forestière dans l'Indochine francaise," *Revue Francaise d'Histoire d'Outre-Mer* 319 (1998); Richard Grove, *Ecology, Climate and Empire: Colonialism and Global Environmental History, 1400–1940* (Cambridge, UK: White Horse Press, 1997); S. Ravi Rajan, *Modernizing Nature: Forestry and Imperial Eco-Development, 1800–1950* (Oxford: Oxford University Press, 2006).

6 This article does not argue that ecology was in fact "apolitical." Instead, it asks why and how ecologists engaged in plantation agriculture attempted to construct an "apolitical" space for their science, often failing to do so.

7 Helen Tilley, *Africa as a Living Laboratory: Empire, Development, and the Problem of Scientific Knowledge, 1870–1950* (Chicago: University of Chicago Press, 2011), 134.

Much ecological knowledge generated for plantations focused on the dominant species of latex producer, *hévéa brasiliensis*. Europeans only gradually became aware of *hévéa* in the Brazilian Amazon even though its latex had long been treated to produce rubber. *Hévéa* is in the family euphorbiaceae, which includes many economically valuable plants such as tapioca.[8] Voon Phin Keong has argued that of all the latex-producing plants, the combination of *hévéa's* ability to regenerate bark and to increase production with more frequent tapping made it the particular favorite of plantations.[9] Many challenges faced the introduction of *hévéa* to Southeast Asia as transplanting it from the Amazon, where it occurs in low densities, to the much more tightly packed plantation monocultures meant that diseases were a constant threat to the industry. During the 1930s, grafted trees began to replace those grown from seeds as labor shortages encouraged plantations to employ fewer tappers working more productive trees. While the majority of latex was harvested in two colonies, British Malaya and the Dutch East Indies, French Indochina also contributed significantly to global supplies of rubber.

This article explores three ways in which plantation agriculture influenced experts as they adapted ecology to colonial Southeast Asia. First, plantation agriculture encouraged experts to generate detailed ecological knowledge that enabled the manipulation of non-human nature. Plant pathologists, for example, employed ecology as they sought to gain mastery over the many diseases of plantation-grown *hévéa*. By contrast, a study of weeding practices shows that when ecological principles were ignored, it was in favor of more visible management of the environment rather than less. Second, plantation agriculture drove experts to assert control over people and their lands through the erasure of alternative forms of environmental knowledge. When experts wrote about native agricultural practices in relation to plantation agriculture, they usually did so in order to disparage these techniques. Third, plantation agriculture prevented experts from seeing ecology as a "subversive" science, one that could challenge the colonial and capitalist projects of reducing human and non-human natures to mere commodities. Even when environmental experts sought to mitigate environmental and social degradation through measures such as soil conservation, they viewed ecology as

8 USDA, NRCS. The PLANTS Database, http://plants.usda.gov (last accessed April 29, 2013). National Plant Data Center, Baton Rouge, LA 70874–4490 USA.

9 See Phin Keong Voon, *Western Rubber Planting Enterprise in Southeast Asia, 1876–1921* (Kuala Lumpur: Penerbit Universiti Malaya, 1976). Biologists are still not certain what purpose the latex serves, but one theory is as a defense against diseases.

a control-oriented science meant to increase production.[10] This article now moves to a case study of French Indochina in order to investigate how ecology and plantation agriculture first became entangled.

Exerting Control Over Non-Human Nature

Curiosity about nature in Southeast Asia predates the formation of colonial states, and missionaries and local scholars have long studied the airs, waters, soils, diseases, flora, and fauna of the region. The earliest textual attention to the flora and fauna of Southeast Asia comes from China where authors viewed the south as a source of spices and medical ingredients.[11] The establishment of the European colonial states in the nineteenth century, however, transformed the study of nature. In French Indochina, a botanic garden established in Saigon in 1864, a mere six years after French military victories over the Nguyen dynasty represented an important step in incorporating the nature of Cochinchina into European knowledge systems.[12] Thirty years

10 Colonial experts left the responsibility for the conservation of flora and fauna to hunters and tourists. See for example, Caroline Ford, "Nature, Culture and Conservation in France and Her Colonies 1840–1940," *Past & Present* 183 (2004); John M. MacKenzie, *The Empire of Nature: Hunting, Conservation, and British Imperialism* (Manchester: Manchester University Press, 1988); Mark Cioc, *The Game of Conservation: International Treaties to Protect the World's Migratory Animals* (Athens: Ohio University Press, 2009); Aline Demay, *Tourisme et colonisation en Indochine, 1898 –1939* (Montréal, Paris: Université de Montréal, Université de Paris 1, 2011).

11 Han Ji and Hui-Lin Li, *Nan-fang ts'ao-mu chuang: A Fourth Century Flora of Southeast Asia: Introduction, Translation, Commentaries* (Hong Kong: Chinese University Press, 1979). My thanks to Michele Thompson for suggesting this source. For an early Western-language publication see João de Loureiro, *Flora Cochinchinensis, Sistens Plantas in Regno Cochinchina Nascentes... Dispositae Secundum Systema Sexuale Linnaeanum* (Ulyssipone: Typis et expensis academicis, 1790). For a comparison with the Atlantic World, see Londa L. Schiebinger, *Plants and Empire: Colonial Bioprospecting in the Atlantic World* (Cambridge, MA: Harvard University Press, 2004); Londa L. Schiebinger and Claudia Swan, eds., *Colonial Botany: Science, Commerce, and Politics in the Early Modern World* (Philadelphia: University of Pennsylvania Press, 2005).

12 In 1865, Jean-Baptiste Louis Pierre (1833–1905) became the director of the garden and one year later the first French-published agricultural and botanical calendar appeared. Between 1881 and 1894, Pierre published a multi-volume treatise as well as articles on particular plants. Louis Pierre, *Flore forestière de la Cochinchine*, 5 vols. (Paris: O. Doin, 1881). See inter alia, *Notice sur une espèce d'Isonandra fournissant un produit similaire à la gutta* (Sai-

later, the colonial government created an experimental field site outside of Bến Cát near the provincial town of Thủ Dầu Một.[13] In 1897, this site became the Ong Yêm experimental station, the first of its kind in French Indochina, and received its first seeds of *hévéa brasiliensis*.[14]

By the mid-1910s, French and Vietnamese colonizers had begun to settle permanently in the midlands of southern French Indochina in order to take advantage of what they saw as the fertility of the red earth region, resulting in a slow but steady growth in the numbers of hectares planted in *hévéa*. The soaring rubber prices of the 1920s led to an explosion in the creation of plantations.[15] By this time many in the colonial government had come to view science as an important adjunct to economic growth and the boom in agricultural production encouraged the formation of new institutions of agricultural knowledge production. One of those pushing for this extension of knowledge was the agronomist Paul Carton.

Carton graduated from the National Institute of Colonial Agriculture in France in 1913 and subsequently joined the Meteorological Service in French Indochina, eventually becoming the head of the Agriculture, Breeding, and Forestry department. In his scientific work, Carton investigated rainfall and sunlight distribution patterns and their effects on plant life, which resulted in a series of articles on the effects of day length and light intensity on plant growth.[16] He also explored how maxima, minima, and optimum conditions

---

gon: Imp. autographique du collège des stagiaires, 1874). See Auguste Chevalier, *Catalogue des plantes du Jardin botanique de Sàigon* (Sàigon: Impr. nouvelle Albert Portail, 1919). See obituary of J. B. L. Pierre in "Miscellaneous notes, XXIII," *Bulletin of Miscellaneous Information (Royal Botanic Gardens, Kew)* (1906): 121–22. Jardin Botanique, *Annales du Jardin botanique et de la Ferme expérimentale des Mares* (Sàigon: Imp. du Gouv. Schroeder, K. et A., 1878).

13 As Hazel Hahn has recently shown, the Saigon Botanic Garden's cultural and touristic functions came to overshadow its scientific work. H. Hazel Hahn, "Botanical Gardens of French Indochina, 1860s–1930s" (paper presented at the The Annual Meeting of the Society for French Historical Studies, Cambridge, MA, April 4–7, 2013).

14 Caresche, "Visite de la station agricole d'Ong Yem et de la ferme-école de Bên Cát," *Groupement sud-indochinois des ingénieurs agricoles, Bulletin de liaison* 4 (1933).

15 Michitake Aso, "The Scientist, the Governor, and the Planter: The Political Economy of Agricultural Knowledge in Indochina During the Creation of a "Science of Rubber", 1900–1940," *East Asian Science, Technology and Society: an International Journal* 3, no. 2/3 (2009).

16 For more on Carton see Institut National d'Agronomie Coloniale, *Anciens élèves de l'Institut National d'Agronomie Coloniale* (Paris, 1929). For more on Carton's theoretical work in ecology see Paul Carton, "Importance des facteurs écologiques 'durée du jour' et 'intensité de la lumière' en agronomie tropicale," *l'Agronomie Coloniale*, no. 183, 184, 185,

influenced the quantity and quality of plant and animal yields. Carton is relevant for the history of ecology because he almost single-handedly adapted the new sub-discipline of écologie agricole (agricultural ecology) to French Indochina after World War I.[17]

Carton's interest in agricultural ecology had roots early in his career. In the late 1910s, Carton joined the editorial staff of the *International Review of the Science and Practice of Agriculture*, published by the Bureau of Agricultural Intelligence and Plant Diseases of the International Institute of Agriculture in Rome.[18] In the 1920, one of Carton's colleagues at the Institute, the Italian Girolamo Azzi, coined the term agricultural ecology.[19] Carton used the educational system in French Indochina as a conduit to introduce his colleague's ideas. In 1918 the colonial government founded the Practical School of Agriculture (*l'Ecole pratique d'Agriculture*, PSA) near the Ong Yêm experimental station in the south and the Higher School of Agriculture and Forestry of Indochina (ESASI) near the botanic garden in the north.[20] Both the PSA and ESASI were meant to train technicians for government services who could improve rural agricultural techniques. When the ESASI was reorganized in 1925, Carton included agricultural ecology in the school's curriculum. Reflecting Carton's expertise, this unit on ecology was included under the heading agricultural meteorology (*météorologie agricole*) and consisted of

186 (1933); —, "Considération sur l'action de la lumière sur les plantes," *Bulletin générale de l'Instruction publique* 10, no. juin-août (1934).

17 Throughout the colonial period in French Indochina, the study of ecology remained an idiosyncratic affair and the term remained absent from many places where it might have been found. The definitive multi-volume treatise on the flora of Indochina, for instance, does not use the word ecology. Although these texts are taxonomic rather than predictive in their orientation, it is not unreasonable to expect to see the term in an era when plant ecology had become an established discipline at many universities in the U.S. and Europe. See Henri Lecomte, Henri Humbert, and François Gagnepain, eds., *Flore générale de l'Indo-Chine*, 7 vols. (Paris: Masson et Cie, 1907–1951),. http://biodiversitylibrary.org/search.aspx?SearchTerm=l%27indo-chine&SearchCat=, (last accessed April 29, 2013 on the Biodiversity Heritage Library).

18 See for example Bureau of Agricultural Intelligence and Plant Diseases, *International Review of the Science and Practice of Agriculture* X, no. 1 (1919).

19 In 1921 Azzi held the first chair in Italy in agricultural ecology. In reviewing Azzi's research, the Royal Academy of the Lynxes (*Reale Accademia dei Lincei*) remarked on how the science of plant ecology "could take on an eminently practical character when it considered the plant grown by man for his needs." Girolamo Azzi, *Ecologie agricole* (Paris: J.-B. Baillière et fils, 1954), 5–6.

20 Michitake Aso, "Profits or People? Rubber plantations and everyday technology in rural Indochina," *Modern Asian Studies* 46, no. 1 (2012).

30 lessons of theoretical instruction and a 10-lesson practicum during the general curriculum of the first year. In these lessons, students examined the influence of the weather and climate on plants and animals and considered questions of plant varieties, acclimatization, and genetics.[21]

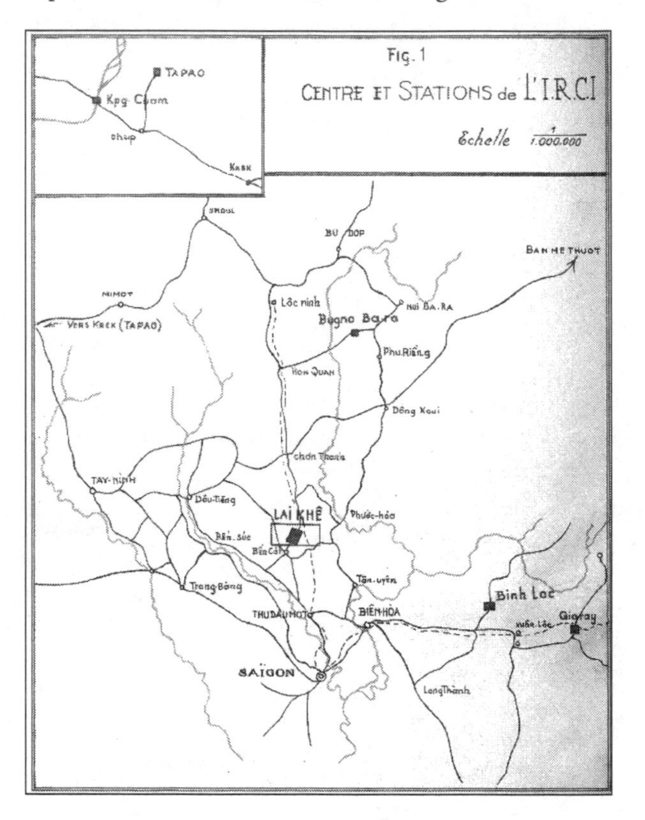

Map of the research station network in the plantation region. Source: L'Institut des recherches sur le caoutchouc en Indochine, Saigon: S.I.L.I., 1942. I would like to thank Philippe Le Failler at the EFEO in Hanoi for his help in locating a pdf version of this source.

21 Paul Carton, "L'Ecologie. Importance de son enseignement dans les écoles supérieures d'agriculture comme «introduction» à divers cours spéciaux. Importance particulière du point de vue colonial," *L'Agronomie Coloniale* 25, no. 227 (1936). For a summary of another article by Paul Carton in *Feuille mensuelle de Renseignements de l'IGA* see Anonymous, «La météorologie agricole en Indochine,» *L'Eveil économique de l'Indochine* 12, no. 566 (1928). This summary depicted climate as the average plus the extremes of weather, i.e. a statistical phenomenon. There was no discussion of culture or of "native" agriculture. It mentioned the importance of large plantations for extending the meteorological network.

The cultivation of *hévéa* also encouraged the production and dissemination of certain kinds of ecological knowledge. Carton pushed for laboratories and plantation-funded agricultural stations to act synergistically and "the station," he wrote, "is the natural extension of the laboratory."[22] Furthermore, by the late 1920s, rubber planters from small and medium owners to the largest multinational companies such as Michelin were taking note of the climate knowledge produced by Carton and others in the laboratory and on experimental fields. Previously such knowledge had been considered esoteric and rather useless but an economic journal from the late 1920s noted that "agriculturalists have begun to notice that the study of atmospheric phenomenon, which influence the development of plants, can be of interest to them and that it is not a priori a waste of time to follow such work."[23]

The synergy between ecology and agriculture gained visibility outside the colony as well. Organizations such as the Parisian group *L'Association Colonies Sciences* (ACS), established in 1925, brought ecological research done in French Indochina to the attention of scientists in the metropole. The founders of the ACS included Auguste Chevalier, Albert Calmette, and other colonial technicians, administrators and members of the colonial economy and political lobby who viewed ecology as useful for economic development (Girolamo Azzi was later an honorary member). The ACS was partially funded by large banks and companies and not surprisingly rubber planters subscribed to this group. The ACS's two purposes according to the *Bulletin du syndicat des planteurs de caoutchouc de l'Indochine* (*BSPCI*) were: first, to put in contact those related to the "development of colonial soil"; and second, to be a documentation center "on the plant and animal products of our colonies." In order to better coordinate efforts between imperial and local governments, the ACS organized conferences, published information, and formed two sub-committees, one to collect soil samples and produce maps and one to study colonial plant parasites and diseases.[24]

Inter-imperial interactions also took place and although those in the rubber industry in the French and British colonial empires (along with the

22 Paul Carton, "L'œuvre de l'institut des recherches agronomiques et forestières de l'Indochine au cours de la période 1925–1943," *Agronomie Tropicale* 1, no. 3–4 (1946): 119.

23 Anonymous, "Information diverses, La météorologie agricole," *L'Eveil économique de l'Indochine* 12, no. 568 (1928).

24 Lettre de l'Association 'Colonies et Sciences', *Bulletin du syndicat des planteurs de caoutchouc de l'Indochine* 90 (1926): 363–64. See also letter from *ASC* asking for financial contributions in "Rapport sur le fonctionnement de l'Association *Colonies-Sciences* en 1926», *Bulletin du syndicat des planteurs de caoutchouc de l'Indochine* 100 (1927): 95.

American commercial empire) often viewed each other as competitors, they also sought to cooperate, especially through the exchange of scientific knowledge. Research in ecology in the British colonial empire both contrasted with and resembled that carried out in French Indochina. While during the late 1920s and 1930s, investigators in British colonial Africa established "the roots of … a 'farmer first' mindset," work on plant pathology related to rubber production confirms the state- and business-centered character of ecology in British colonial Southeast Asia.[25] Arnold Sharples, who worked as a mycologist in the Department of Agriculture in Malaya beginning in 1913 as well as in the Pathological Division of the Rubber Research Institute, offered his thoughts on the profitability of science. Similar to his French counterparts, Arnold drew a distinction between applied and pure science, he wrote that "in the former, profits must be clearly visible, while, in the latter, the completion of a line of research work is the primary consideration, and whether or not profitable issues may be the ultimate outcome does not affect the question."[26]

While dividing scientific work between "applied" and "pure" may appear to leave room for science driven by motives other than profit, Sharples underscored the business aspects of plant pathology when he wrote: "I believe that all plant pathologists who intend undertaking applied research in large agricultural combines should have a knowledge of business and financial details included in their training curriculum."[27] The concern of the ACS about plant pathology and Sharples' own research interests suggest the importance of pathology for the profitable operation of monoculture plantations. This section now considers how colonial experts applied control-oriented ecology to the pathologies of *hévéa*.

Plant Pathology

The study of *hévéa* diseases on plantations illustrates how colonial experts attempted to use ecological knowledge to exert control over non-human nature. Experts in the British colonies contributed to this international effort and in 1911, Thomas Petch published one of the major early works on

25 Tilley, *Africa as a Living Laboratory.*

26 Arnold Sharples, *Diseases and pests of the rubber tree* (London: Macmillan and Co., ltd., 1936), 2.

27 Ibid., 3.

hévéa biology on plantations. A decade later, Petch expanded a section in the original monograph to a standalone volume dedicated exclusively to *hévéa* pathologies called *The Diseases And Pests Of The Rubber Tree*. Petch noted the rapid expansion in knowledge about pathologies, including information on South American Leaf Disease (later renamed South American Leaf Blight).

The form and content of *Diseases* is reminiscent of a medical text on human ailments and the book is divided up by the part of the tree under attack, including root, leaf, and stem, as well as by the particular pathologies, the symptoms of which Petch graphically described, and their causes. Petch's introductory chapter on general sanitation offers further parallels with medical approaches to human diseases. Petch wrote of the "visitations" that characterized previous understandings of plant diseases and of how only recently experts had realized "that plant diseases are as inevitable as those of men and animals."[28]

Petch advocated prevention over remediation and he argued "the pathologist should be consulted beforehand, not five or six years afterwards when some disease has already appeared." These preventative measures involved general sanitation and Petch warned against over-crowding of *hévéa*, which would lead to diminished rubber yields, both through stress and the easy communication of disease. Initially, plantation practice had been to plant trees from 100 to 200 per acre up to 500 per hectare and Petch spoke of thinning trees to 150 per hectare or less. He also emphasized the dangers of "jungle stumps," which were tree stumps left to rot because removal was expensive. During this process these stumps became carriers of many of the diseases that then spread to *hévéa*, either through fructification, the spread of spores, or through fungus hyphae and direct root contact.[29]

Agricultural scientists looked to their colleagues in the medical sciences for models of how to approach plant pathologies and once diseases did strike, the steps recommended often resembled coercive public health measures. If a disease was minor and, more importantly, did not threaten to spread to other trees, then "tree surgery" could be performed, i.e. the diseased part could be cut out and the resulting wound treated. If, however, diseases threatened to be epidemic more drastic measures were required. The treatment for *fomes lignosus*, a pathology earlier known as *fomes semitostus*, offers an example of

28 Thomas Petch, *The physiology & diseases of Hevea brasiliensis, the premier plantation rubber tree* (London: Dulau & Co., limited, 1911). —, *The diseases and pests of the rubber tree* (London: Macmillan, 1921), 1.

29 Petch, *The diseases and pests of the rubber tree*, 2.

drastic measures. *Fomes* was one of the most feared root diseases of Southeast Asia because it moved quickly and could spread widely before above ground signs signaled its existence. Not only were the visibly diseased trees dug up and their roots removed but planters had also to identify apparently healthy trees that had been infected. If the *fomes* had spread enough, then infected ground was covered in lime and forked, mixing in the lime. In addition, some experts recommended digging a trench to almost a meter deep and covering it in lime. These measures of controlling space and individual organisms were analogous to steps taken to quarantine human communities during outbreaks of plague or cholera.

After Petch, Dr. Alfred Steinmann published the next major text on rubber diseases in 1925 in Dutch, later translated into English in 1927. Together, Petch's and Steinmann's monographs were viewed as the industry standards.[30] Steinmann discussed diseases caused by identifiable parasites but he also noted that *hévéa* could suffer from environmental stressors. He mentioned in particular the harm to trees that could result from worn-out soils, their nutrients leeched by tropical rains and by previous crops. Other damage to the tree could result from overexploitation of *hévéa*'s bark. The solutions for these diseases varied from application of pesticides such as Paris Green to the use of fertilizers to restore soil fertility. Yet, to a certain extent diseases were viewed as unavoidable and Emile de Wildeman, a Belgian expert on rubber, wrote that diseases inevitably appear on monoculture plantations.[31]

How did planters put this theoretical knowledge into practice? The scientific literature suggests that managers and researchers began with general knowledge about the environment and moved to specific plantation diseases. Especially when agronomists could not identify individual pathogens as causing stress on the trees, they drew on knowledge of climates, soils, flora, and fauna, and their connections in order to reason about the maladies of monocultures. Yet, plantations also provided productive places for researchers to theorize about plant diseases and major advances in the study of *hévéa* pathologies were made at the beginning of the 1920s, after rubber had been grown on plantations in Southeast Asia for more than a decade. Although Auguste Chevalier stated that much was still unknown about *hévéa*'s biol-

30 Alfred Steinmann, *De ziekten en plagen van Hevea brasiliensis in Nederlandsch-Indië* (Buitenzorg: Archipel Drukkerij, 1925). See preface of Sharples, *Diseases and pests of the rubber tree.*

31 E. de Wildeman, "Plantations et maladies de l'hévéa," *Revue de Botanique Appliquée et d'Agriculture Coloniale* 6, no. 53 (1926): 19.

ogy, his report on the 1921 International Exposition on Rubber and Other Tropical Products in London listed 18 known diseases of *hévéa.*[32]

By the mid-1930s, yet more was known about the diseases of *hévéa* but there still remained confusion in the field, a state that threatened the relationship between planters and ecology experts. Arnold Sharples decried the inconsistencies in naming that had cropped up in the field of plant pathology. For example, Sharples discussed a debate between Petch and Van Overeem over the naming of *fomes lignosus*, also known as White-Root disease. According to Petch's 1921 publication, *fomes lignosus* was the same as the existing pathology called *fomes semitostus*. In 1923, however, Van Overeem challenged this classification and proposed the name *Rigidoporus microporus*, Swartz. Van O., placing *fomes lignosus* in a more general class of fungus. Yet Sharples seemed less interested in the minutia of scientific controversies than in finding practical solutions to problems. "Little can be gained," Sharples lamented, "by constant disagreement over the correct systematic position and name of an organism causing a common plant disease, and it certainly leads to lack of confidence between planters and the pathological workers (sic) who are called upon for advice."[33]

Cleaning Weeding

While each of the steps needed to create plantations including clearing of forests, planting of saplings, tapping of trees, and regenerating bark had implications for disease control, clean weeding received much attention. At stake in discussions over proper weeding regimes were questions of ecological knowledge on the one hand and aesthetics on the other. Such debates show that disagreements existed over modernity and the most effective environmental practices on plantations.[34]

32 Auguste Chevalier, "Situation de la production du Caoutchouc en 1921," *Revue de Botanique Appliquée et d'Agriculture Coloniale* 1, no. 2 (1921): 78; ———, "Rapport sur la 5e Exposition internationale du Caoutchouc et des autres Produits tropicaux à Londres," *Revue de Botanique Appliquée et d'Agriculture Coloniale* 1, no. 2 (1921): 340–42.

33 Sharples, *Diseases and pests of the rubber tree*, 4.

34 Barlow has noted the interest of the cleaning weeding controversy for students of technology. Colin Barlow, *The Natural Rubber Industry: Its Development, Technology, and Economy in Malaysia* (New York: Oxford University Press, 1978). See chapter 4, "The Technologies of Production", 112–159.

On the one hand, experts who advocated for clean weeding, the elimination of all unwanted plants from between the rows of *hévéa* trees, called upon ecology to bolster their position. Some argued that by taking out the "weeds" that could later turn into combustible material, the risk of fire was greatly reduced. Other experts believed that superfluous vegetation could take away from *hévéa* the "force de végétation," a vague concept that held that environmental conditions could only support so much vegetable matter. This concept is analogous to the idea of "carrying capacity" in which given environmental conditions of water, light, and space can sustain a certain amount of life. In the case of plantations, *tranh* (Lalang grass) was seen as stealing nutrients from *hévéa*, and various methods were recommended for clearing weeds including hoeing and arsenate.[35]

On the other hand, the opponents of clean weeding also used science to bolster their positions. H. N. Ridley of the Singapore botanic gardens advocated leaving some undergrowth as a means of retaining moisture and for the purposes of green manure. Yves-Marie Henry (1875–1966) argued that clean weeding resulted in the loss of humus.[36] While most planters in French Indochina initially adopted clean weeding practices, many eventually decided to leave plant growth, at least until the crown of the *hévéa* trees shaded this undergrowth out. But how did planters decide between these systems and what questions other than environment were at play?

Unlike research on diseases, where planters and experts worked to close the distance between the field and the laboratory, planters often ignored the writings of experts on clean weeding, preferring aesthetic criteria to sound reasoning about the environment. Christophe Bonneuil has shown how aesthetic reasons continued to trump other considerations and even agricultural

35 Weeding regimes could have other environmental consequences and industrial agriculture and transnational capital had to deal with issues such as erosion. "Actes officiels: Circulaire au sujet de l'obligation pour les acquéreurs à titre onéreux de terrains domaniaux de borner leurs propriétés avec des matériaux de longue durée" and "Renseignements à l'usage des Planteurs de Caoutchouc (suite)," *Association des planteurs de caoutchouc de l'Indochine* 11 (1911): 79, 81. See also Piers M. Blaikie, *The Political Economy of Soil Erosion in Developing Countries* (London: Longman, 1985). More recently see David Anderson, *Eroding the Commons: The Politics of Ecology in Baringo, Kenya, 1890s–1963* (Athens: Ohio University Press, 2002); Kate Barger Showers, *Imperial Gullies: Soil Erosion and Conservation in Lesotho* (Athens: Ohio University Press, 2005).

36 Anonymous, "Renseignements à l'usage des Planteurs de Caoutchouc, Visite de M. H. N. Ridley", *Association des planteurs de caoutchouc de l'Indochine* 21 (1911): 613. Yves-Marie Henry, *Terres rouges et terres noires basaltiques d'Indochine; leur mise en culture* (Hanoi: IDEO, 1931), 73.

scientists themselves praised planters for keeping a neat plantation appearance, regardless of the cost or benefit for latex production.[37]

In the British colonial empire, Arnold Sharples argued for a balanced approach to clean weeding. On some types of grounds including flat and rather swampy grounds, this system could be useful, Sharples stated, but for undulating inland hills, clean weeding should be avoided because of erosion problems. His take on clean weeding fits into his overall approach to the application of the principles of "forestry methods" to plantations. These methods, Sharples argued, presented three main advantages: (1) shading soil for temperature reasons; (2) preventing soil erosion; and (3) providing humus. Sharples argued that only "natural" undergrowth was to be allowed as other types of undergrowth could lead to increased diseases. In any case, Sharples noted the tension between scientific methods and those more commonly used among planters.[38]

The slippage between ecology and morality when discussing well-weeded spaces can be seen in a 1933 report about the Canque plantation in Indochina. R. Caty, an alumnus of the agricultural school in Grignon and head of the genetic laboratory of the Indochinese Bureau of Rice (OIR) at the time, wrote that Canque was chosen for its visually perfect presentation, including its clean weeding, which gave the observer the impression that the owner cared for his operation. While Canque's old tree stock gave only modest yields of 507 kilograms per hectare, it seems that Caty wished to emphasize that even during the economic crisis, good maintenance and appearance were necessary.[39] In other words, the desire for legibility, the act of simplifying in order to effect intervention, extended beyond state projects to small businesses as well.[40]

37 For a detailed analysis of clean weeding, see chapter 5 in Bonneuil, *Mettre en ordre et discipliner les tropiques*.

38 Sharples, *Diseases and pests of the rubber tree*, x, 454.

39 R. Caty, "Visit de la Plantation de Canque," *Groupement sud-indochinois des ingénieurs agricoles, Bulletin de liaison* 4 (1933). See also ———, "L'amélioration des plantes de culture indigène aux colonies," *l'Agronomie Coloniale* 25, no. 218, 219 (1936).

40 A classic state on legibility projects is James Scott, *Seeing Like a State: How Certain Schemes to Improve the Human Condition Have Failed* (New Haven: Yale University Press, 1998).

Exerting Control Over People and Their Lands

Central to the colonial project were the ideological ways in which various actors deployed ecology in an attempt to control people and land. More specifically, a post-World War I concern among officials, experts, and planters about native agriculture, whose defining elements were more socio-economic than technical or botanical, grew from a shared desire to assert authority over these practices. Colonial officials attempted to legitimize their control of French Indochina by delimiting spaces of "natural resources," such as rubber plantations, and removing culture from nature. Colonial planters spoke the language of ecology when it served their need to appropriate land for their plantations from previous inhabitants. Finally, colonial experts attempted to speak for nature by claiming that they alone, as scientists, knew how best to manage the environment, either for conservation or as a resource. Each of these colonial actors employed ecology as a tool to critique "indigenous" agricultural practices and thus delegitimize "native" claims to the land.

In 1925, Yves Henry wrote that "studies [on native agriculture] have rarely led to a practical plan for progressive improvement mainly because they have not been consistent and coordinated, nor always submitted to scientific discipline. This discipline alone," Henry continued, "enables a unity of doctrine, which is the source of all productive action."[41] Henry was one expert who harbored a growing interest in local knowledge that could be termed "vernacular science."[42] Yet, native agricultural knowledge continued to be assigned an inferior place among agricultural scientists. While planters and experts attempted to correct their own misconceptions about the environment, and sometimes expressed concern for native welfare, they viewed native agriculture as a threat to plantations and either ignored or denigrated these practices. This trend was only exacerbated by the further privatization of knowledge production and not until the 1930s did a few human geographers begin to take seriously the techniques of indigenous farmers.

Interest in native agriculture among officials, experts, and planters coalesced around the relationship between "indigenous" and "European" agriculture. Fire, in particular, was valued differently whether it was used to clear land for plantations or for swidden. Plantation owners, who coveted swidden land, condemned native use of fire, while often hiring these same people to

41 Quotes from report of Yves Henry, 23 March 1925 in Carton, "L'œuvre de l'institut des recherches agronomiques et forestières de l'Indochine", 116.

42 Tilley, *Africa as a Living Laboratory*.

clear land with swidden techniques. Many colonial experts, who viewed fire as destructive of ecosystems, supported a crusade against swidden agriculture even as they chastised plantations for equally destructive practices. Finally, colonial officials, who attempted to manage society, adopted an ambivalent position and at times spoke of the necessity of fire for native lifestyles.

Clearing Fires

Fire had long existed in eastern Cochinchina but the creation of plantations changed the region's fire regime and growing land values resulted in increased pressure to suppress fire.[43] The extended dry season meant that fires were fairly common and was used by uplands peoples in swidden agriculture. Swidden, called *rẩy* in Vietnamese, is a farming practice that clears patches of forests with fire in a rotating system that cycles through several years. After swidden became a topic of scientific interest at the end of the nineteenth century, early studies tended to dismiss its viability and blamed low numbers of Montagnards in the central highlands partly on this practice. The geographer Mark Cleary has cited a colonial agronomist who held that "settled farming belonged to a higher plane than shifting cultivation or hunting and gathering."[44] Environmentally, swidden was viewed by many experts as bad for the soil and, largely due to concerns about Cochinchina's forests, swidden was made illegal in 1894.[45]

Fire was not simply a natural event and colonial discourse reinforced the idea of Montagnards and their agriculture as a threat in many planters' and officials' minds even as environmental changes introduced by colonialism itself played a large role in starting and sustaining fires. Although planters later condemned swidden in an attempt to bolster their own land claims and protect their investments, knowledge of this forest clearing technique made upland groups invaluable during the initial stages of plantation creation. After planters oversaw the cutting down of forests and the draining of

43 P. M. Allouard, *Pratique de la lutte contre les feux de brousse* (Hanoi: IDEO, 1937); A. Consigny, *Considérations sur les feux de brousse leur méfaits et la possibilité de les enrayer* (n.l., n.d.).

44 Mark Cleary, "Managing the Forest in Colonial Indochina c. 1900–1940," *Modern Asian Studies* 39, no. 2 (2005): 279. Fire seems to have been an important part of some Montagnard societies, as the studies of Georges Condominas and Gerald Hickey have shown.

45 For discussions among foresters in the French and British empires, see Rajan, *Modernizing nature*.

marshes, fires began to burn hotter and faster as the detritus of *tranh* and felled tress and the monocultures of *hévéa* all provided excellent fuel. Cleary quotes a head forest officer who argued that "the damage from burning was due as much to plantation owners and their workers … as to the indigenous groups."[46]

In the early twentieth century, swidden continued as there was little French presence in the midland and upland regions and most Montagnards were unaware of the new legal status of the land they occupied. Some French colonial officials recognized the necessity of swidden as a source of food for the uplands peoples and the fact that Montagnards were ignorant of French laws. For example in a 1904 letter to the district head of the Cochinchinese Forest service, the temporary head of the Indochinese Forest Service, Chapotte, discussed the attempts to curtail swidden. He argued against the strict application of punishments for this practice, especially the seizing of the harvest, as such punishments only served to drive Montagnards away from the French outposts and further into inaccessible forest reserves. Instead, Chapotte suggested that "the solution that the Goucoch [Governor of Cochinchina] recommends and judges to be reasonable, that is to say the fixing of the Moïs [a derogatory word for those living in the midlands and uplands] on the land, is certainly the solution that best meets our humanitarian ideas and our peaceful penetration." Chapotte even suggested an agricultural system current in parts of France as a possible model for settling Montagnards and preserving forest reserves.[47]

The spread of plantation agriculture, however, introduced a hard edge to colonial discourse about Montagnards and their use of fire. On the one hand, planters needed Montagnard populations and their skills as forest clearers, guards, and overseers. Partly because of the importance of their labor, partly because of their picturesque value, the planters' journal the *BSPCI* featured a "moïs" cover in the early 1910s. Montagnard labor maintained more than symbolic value into the late 1920s and in 1928 the *Société Agricole et Industri-*

46 Cleary, "Managing the Forest in Colonial Indochina c. 1900–1940", 274.

47 TTLT 1 663 AFC, Répression des délits de "rãys" constatés en pays Moïs 1904. M. Chapotte, Inspecteur Adjoint Eaux et Forêts, chef p.i. du service forestier de l'Indochine à M. chef de la circonscription forestière de Cochinchine, Hanoi, 7 sept 1904. Quote from page 2.

elle de Cam-Tiên listed "moïs" and Cham among its 600 "free" laborers, i.e. those without a contract.[48]

On the other hand, those advocating industrial agriculture condemned swidden. In 1911, for example, the planter M. Sipière discussed the dangers of the fires that "moïs" used to follow their "lazy" way of agriculture. He admitted that these fires helped clear *tranh* and provided range for large mammals to roam and to be hunted; but Montagnard fires, argued Sipière, threatened the plantations.[49] A similar situation for the central highlands was noted in a 1932 report from the Annam Chamber of Commerce and Agriculture. The report's author was troubled by the fact that fires "lit in successive lines and sometimes over a distant of more than 2 km by 100s of moys [sic], are above all practiced in the regions where French colonization has been settled and that the other regions are little touched by the fires."[50] This author did not note the irony that French colonization itself was a disturbing presence.

During the rubber boom in Indochina of the 1920s, the question of clearing for plantations generated two schools of thought, which Christophe Bonneuil has nicely summarized. The first method, used on the Suzannah plantation in Cochinchina for instance, consisted of removing stumps mechanically and completely clearing the land of *tranh*. This method, called "culture intensive," required high initial inputs of non-human energy such as tractors or, more commonly, large herds of cattle, at times almost one buffalo per hectare of *hévéa*. Suzannah and Anloc in 1920 had 2200 cattle for 2700 hectare planted. Cattle both provided the necessary power to clear the land and the means to fertilize it with manure. They also had many requirements, including grass on the ground and veterinary attention to deal with the epizootics that ravaged the ungulates of the region.[51]

The second method, called *méthode malaise*, mimicked swidden and consisted of cutting trees and then burning the ground without taking out the

48 TTLT 2 Goucoch IIB.56/029, Procès Verbal de Visite de la Plantation de Cam-Tien, 1928 janvier 14.

49 "Ray.–Dangers d'incendie", *Association des planteurs de caoutchouc de l'Indochine* 22 (1911): 641.

50 TTLT 4 3113, Vœux de chambre mixte de commerce et d'agriculture de l'Annam, Touraine, 1932.

51 Girard, director of the Suzannah and Anloc plantations, was an apostle (apôtre) for this method. For more on the two schools of thought regarding plantation clearing, see Bonneuil, *Mettre en ordre et discipliner les tropiques*, 303. See also the 1917 unpublished manuscript of George Vernet, who directed Dr. Yersin's plantation in Nha Trang, TTLT 2, TDBNV N5.59.

stumps. This system required intensive labor during the planting phase as workers used pickaxes to remove the root systems of *tranh*, which fire did not eliminate. Used in Ceylon (now Sri Lanka) and imported by planters such as the Belgian financier and plantation owner Adrien Hallet (1867–1925), this second system was employed on the fertile red earth soils and allowed for a more exclusive devotion of resources to latex production. After the clearing of the land with fire, soils were exposed to the climate and chances of erosion and laterization, or the formation of a stone-like substance, increased. Although some disparagingly compared the method of "our neighbors in the East [the British]" to the method of the "moïs," this system won out during standardization of plantation practices in French Indochina during the 1920s.[52]

Throughout the 1930s, planters continued to ignore the correspondence between clearing practices and swidden and officials often depicted swidden as irresponsible and destructive of French economic activity. In 1931, Yves Henry, who had transferred from his post in Africa to serve as Head of Agriculture, Breeding, and Forestry (AEF), condemned swidden as destructive of the soil.[53] "This system of cultivation," he wrote, "has the consequence of impoverishment of the soil by continuous cultivation and the *dénudation* that have considerably diminished the value of this land capital." French colonization, according to Henry, had worked to save the soil quality by settling agriculturalists and reserving forests.[54] By contrast, the forestry commissioner Consigny admitted the impossibility of eliminating swidden and acknowledged the source of food that it provided for Montagnards, even as he consistently opposed brush fires whatever the cause.[55]

Only in the 1940s and 1950s did scientific opinion about swidden begin to shift with authors either showing the relatively minor impact of this technique or how it was a rational approach given a sparse population and sloping agricultural land. An exhaustive study of French Indochina's forests published in 1951 for example, held that the open forests of the south were

52 See William Gervase Clarence-Smith, "The Rivaud-Hallet Plantation Group in the Economic Crises of the Inter-War Years," in *Private Enterprise during Economic Crises: Tactics and Strategies*, ed. Pierre Lanthier and Hubert Watelet (New York: LEGAS, 1997).

53 Académie des sciences d'outre-mer, "Hommes et destins: dictionnaire biographique d'outre-mer," Académie des sciences d'outre-mer. Tome V: Yves-Marie Henry (1875–1966), by André Angladette, 245.

54 Henry, *Terres rouges et terres noires basaltiques d'Indochine*, 34.

55 Consigny, *Considérations sur les feux de brousse leur méfaits et la possibilité de les enrayer*, 1, 8.

created by climate and soils and not the human use of fire.[56] Pierre Gourou, a human geographer, also came to this conclusion in his writings, admitting that swidden could be well adapted for certain environments. Gourou did foresee problems if the highlands became more densely populated but, like Consigny, understood the dilemma in trying to eliminate the *rẫy* system.

> The rây is, very often, the sole means of existence of mountain populations; even though the latter are not numerous it is impossible to deprive them of their principle source of food because the râys are ruinous to the future of Indochinese economy; such an attempt would not be humane, and it would be ridiculous because unenforceable. Rây can disappear only if the peoples who practice it are given means of existence by other agricultural methods....Such an attitude, purely negative and of policing nature, would be intolerable.[57]

In his history of the rubber industry in Vietnam, the expert Đăng Văn Vinh went further than Gourou and offered an appreciative account of the bamboo-clearing skills of the Montagnard, termed ethnic minorities in postcolonial Vietnam.[58]

Several scholars have come to view ecology as part of a "subversive" reaction to the ideals of "free-market" capitalism that threatened to annihilate nature. As Tilley has shown for British colonial Africa, a "subversive relationship ... could exist between science and empire, particularly in the era of late European colonialism."[59] But the question remains: In relation to the rubber plantations of colonial Southeast Asia, was ecology viewed as subverting capitalism and colonialism or were the profit motive and proper environmental management seen as mutually supportive? On the one hand, according to the historians Beinart and Hughes, rubber plantations in Brit-

56 Bernard Rollet et al., *Les forets claires au sud-Indochinois: Cambodge, sud-Laos, sud-Viêtnam*, 2 vols. (Saigon: Centre de Recherches Scientifiques et Techniques du Cambodge, Laos, Viet-Nam, 1952).

57 Pierre Gourou, *L'Utilisation du sol en Indochine française* (Paris: P. Hartmann, 1940), 494.

58 Đặng Văn Vinh, *100 năm cao su ở Việt Nam* (TP HCM: NXB Nông Nghiệp, 2000), x. For work in English defending swidden see inter alia, the writing of Gerald Hickey, a U.S. anthropologist doing research in Central Highlands. In "Memorandum for Record: Montagnard Agriculture and Land Tenure" of April 2 1965, Hickey wrote: "Using the swidden method has the advantage of leaving tree roots in the soil which helps to retain the structure. If a plow were used in these circumstances there would be grave danger of having the top soil wash away." From Oscar Salemink, *The Ethnography of Vietnam's Central Highlanders: A Historical Contextualization, 1850–1990* (Honolulu, HI: University of Hawai'i Press, 2003), 243.

59 Tilley, *Africa as a Living Laboratory*, 24.

ish Malaya were relatively environmentally neutral.[60] On the other hand, economic incentives did not prevent environmental degradation and they effectively stymied attempts to protect nature arising from aesthetic considerations, tourism, hunting, and later nationalism. In the case of rubber plantations the bottom line dictated whether experts fought against the destruction of nature or not.[61]

Ecology as Subversive?

While some studies have found the roots of "environmentalism" among colonial experts, these were not the people who promoted the protection of nature in Southeast Asia during the early twentieth century.[62] In the French colonial empire, for example, 13 national parks (*parcs nationaux*) were established between 1923 and 1927 in Alger and Constantine, while none were established in French Indochina. In the late 1930s, the forester Y Marcon explained that preservation of natural diversity had its place but the real job of foresters involved exploitation. "Within this [forestry] framework," Marcon stated, "we leave aside all that relates to the protection of the natural flora, certainly an activity that is very interesting, but one instead attached to sentimental or theoretical considerations, for we are occupied with the economic activities of the Forest Service."[63] This Service, he concluded, was meant to make the most of the forests' products and both conservation and exploitation were directed with commercial ends in mind.

The absence of conservation and preservation efforts happened despite international interest in the nature of colonial Southeast Asia and emphasis

60 William Beinart and Lotte Hughes, *Environment and Empire* (New York: Oxford University Press, 2007).

61 Paul B. Sears, "Ecology—A Subversive Subject," *BioScience* (1964); Karl Polanyi, *The Great Transformation* (Boston: Beacon Press, 1944); David Henley, "Natural Resource Management and Mismanagement: Observations from Southeast Asian Agricultural History," in *A History of Natural Resources in Asia: The Wealth of Nature*, ed. Greg Bankoff and P. Boomgaard (Basingstoke: Palgrave Macmillan, 2007).

62 Richard Grove, *Green Imperialism: Colonial Expansion, Tropical Island Edens and the Origins of Environmentalism, 1600–1860* (Cambridge: Cambridge University Press, 1995).

63 Frederic Thomas, *Histoire du regime et des services forestiers français en Indochine de 1862 à 1945* (Hanoi: The Gioi, 1999), 69. Quoted from Y Marcon, *Le role des services forestiers aux colonies*. Actes de comptes-rendus de *l'Association colonies sciences*, 1938, 2.

was placed instead on protecting sites of cultural attainment and ancient civilizations.[64] Factors including tourism, the supposed fecundity of tropical nature, ideas about the ancient civilizations of the East, and the strength of scholarly bodies such as the *Ecole française d'Extrême-Orient* (EFEO) and the Malayan Branch of the Royal Asiatic Society (MBRAS) all played a role. So too did the fact that most experts viewed ecology through an economic lens, which meant that they focused on manipulating biophysical environments in order to increase the yield of plants and animals that were produced in, and productive of, these environments. Plantation practices in particular were aided by research on specific aspects of the environment such as soils. While this research did not mention ecology, it often took a synthetic approach to the environment and looked at interactions among humans, plants, and soils. Significantly, these studies began to create a new appreciation of tropical soils among Western readers. Rather than the wild fertility that seems to lead to humid jungles, tropical soils came to be seen as quite fragile, an awareness that helped to transform scientific attitudes concerning questions about clean weeding and swidden.

In French Indochina, the first attempt to gather scattered soil analyses was undertaken by Yves Henry. In his 1931 monograph he argued that most peoples' perceptions of the soils of the *terres rouges*, the main plantation region, were either based on ignorance or influenced by financial speculation.[65] Newly constructed roads and recently drawn maps enabled Henry to offer an understanding of soils that was at the same time more comprehensive and more detailed. He directed his writing at both planters and agricultural scientists with the dual goals of helping planters and advancing knowledge about important types of soils. Henry's study was representative, if ambitious, for its time in two respects. First, he attempted to describe soils, climates, and vegetation in their interactions as a whole. Second, he sought to ensure the profitability of plantations.

To take one of Henry's examples, the soils of the Kompongcham-Mimot massif in colonial Cambodia were rich in nutrients and were fairly resistant against laterization. But with the forest cover removed, followed by the practice of clean weeding, Henry warned, these soils could rapidly lose their nutrients and humus. Theories about desiccation that linked forest clear-

64 Ford, "Nature, Culture and Conservation in France and Her Colonies 1840–1940." See also papers presented at the Premier Congrès international pour la protection de la nature, faune et flore, sites et monuments naturels, Paris, France, 31 mai–2 juin 1923, 355–56.

65 Henry, *Terres rouges et terres noires basaltiques d'Indochine*, 7.

ing to changes in regional climates also influenced Henry's thinking and he noted the drastic reduction in rainfall (up to 50 per cent) in some places. Henry showed for instance how rainfall levels on Mimot, a plantation on Kompongcham-Mimot massif, went from 3256 millimeters in 1927 right after the beginning of deforestation to 1638 millimeters in 1929 at the end of this process.[66]

As mentioned in the case of swidden, by the late 1930s the attention paid to the environment promoted by, and useful to, rubber plantations had led to critiques of inappropriate agricultural practices. As experts gained a better appreciation of local conditions, some even valorized "indigenous" farming techniques, a stance that could be construed as anti-colonial. Furthermore, many officials viewed rubber plantations with ambivalence based on a concern about lack of state control over these spaces. In this way, ecology could unexpectedly challenge the operation of rubber plantations.[67]

But even as ecology began to take into account native knowledge, it retained in Southeast Asia its supportive stance towards the colonial mission. In the late 1930s, Paul Carton reaffirmed the economic usefulness of agricultural ecology, offering an extended argument in his 136-page textbook for a first-year ecology course at the ESASI.[68] Published in 1938, this textbook highlighted both the economic usefulness of ecology for increasing plant and animal yields and the use of ecology for helping humans to adapt to new climates and to resist diseases. As in the 1920s, Carton defined the purpose of agricultural ecology as discovering the optimal conditions for crop growth. "Humans," he wrote, "try to get a maximum yield, in both quality and quantity, of products necessary for their needs from the plants that they exploit or cultivate."[69]

66 Henry, *Terres rouges et terres noires basaltiques d'Indochine*, 44. For earlier discussions of desiccation see Grove, *Green Imperialism*.

67 See for example, Justin Godart, *Rapport de mission en Indochine, 1er janvier—14 mars 1937* (Paris: L'Harmattan, 1994). In his report, the former minister of health wrote both approvingly and with misgivings about plantations as "forests without birds".

68 Paul Carton, *Cours de Climatologie et d'Ecologie* (Hanoi: Impr. Ngo-Tu-Ha, 1938), 1.

69 Ibid., 3.

Conclusion

Not until after 1945 did colonial experts show high levels of interest in "local" knowledge, both a result of fading colonial states and the decolonization process. Some human geographers, agronomists, and agricultural agents and engineers even published accounts of native agriculture that were highly positive. In this way, French, British, and "native" researchers were not deficient in a critical approach to the production of environmental knowledge and during the interwar period there existed a "pervasiveness of criticisms generated from within by experts active in the field."[70] In colonial Africa, this critical spirit was motivated both by a desire to improve the fit between theory and reality and by a concern about the effects of colonial interventions. In Southeast Asia, however, there was never an equivalent to the African Research Survey and its combination of social anthropology and environmental sciences. The lack of such an epistemological framework meant that "scientific" plantation agriculture continued to be more highly valued than "native" practices.

The analysis of the effects that rubber plantations had on knowledge about the environments in Southeast Asia suggests that plantation agriculture played a decisive role in making ecology a business and state oriented science. This conception may be surprising at a time when ecology, especially in the tropics, can call to mind scientists studying the "wild profusion" of animals, birds, and plants.[71] The use of ecology to critique imperial and neo-imperial projects only found its stride during the Vietnam War era when the spraying of Agent Orange encouraged ecologists and biologists to adopt "subversive" attitudes towards such attempts to control environments.[72] More recently, changes in land use cover and carbon dioxide production and sequestration have become standard concerns in studies of plantation agriculture.[73] Only during the second half of the twentieth century were variants

70 Tilley, *Africa as a Living Laboratory*, 23.

71 Celia Lowe, *Wild Profusion: Biodiversity Conservation in an Indonesian Archipelago* (Princeton, NJ: Princeton University Press, 2006).

72 David Zierler, *The Invention of Ecocide: Agent Orange, Vietnam, and the Scientists Who Changed the Way We Think About the Environment* (Athens, GA: University of Georgia Press, 2011); Edwin A. Martini, *Agent Orange: History, Science, and the Politics of Uncertainty* (Amherst, MA: University of Massachusetts Press, 2012).

73 For an early example, see John F. Richards and Elizabeth P. Flint, "A Century of Land-Use Change in South and Southeast Asia," in *Effects of land-use change on atmospheric CO2*

of ecology that had been central to imperial projects used to question the efficacy of colonial methods and even reject the civilizing mission altogether.

concentrations: South and Southeast Asia as a case study, ed. Virginia H. Dale (New York: Springer-Verlag, 1994).

Apple Orchards in Southern Brazil: An Environmental History

Jó Klanovicz

The expansion of modern commercial apple orchard planting (*Malus domestica*) in Southern Brazil occurred after 1962 in the city of Fraiburgo, Santa Catarina state, based on the capitalist intensification of Brazilian agriculture in the second half of the 20th century. Planting was made on areas which were previously covered with Ombrophyllous Mixed Forest (OMF), more particularly with Araucaria pine trees (*Araucaria angustifolia*).[1]

Apple planting in Fraiburgo would have two basic features: adaptation and tree growing in different ecological conditions during this period, and the existence of a market for these very cultivars. While in the first half of the 20th century such agricultural conditioning had been markedly dictated by the spreading of tropical products in European countries as well as in the USA—devastating large tropical forest regions in order to build homogeneous plant landscapes for example of sugar cane and coffee. However, the construction of peripheral so-called "tropical" markets such as Brazil also served to expand the fruit market, but these fruits were also from temperate climates. Since Brazil copied the *American way of life* particularly after the end of World War II, the capitalist intensification of agriculture also sug-

1 The Ombrophilous Mixed Forest is part of the Brazilian Atlantic Forest biome and had been the richest forest in timber in Southern Brazil between the end of the 19th century and the 1960s. From the 200,000km^2 of forest spread throughout the southern states of the country, Santa Catarina held approximately 50,000km^2. About 25 to 37 millions of commercial Araucaria pine trees (40cm of trunk diameter or more) were extracted between the years of 1910 and 1940 only in the state's mid-western region. In March 2010 the project "Dominated Nature: occupation and deforestation of Western Santa Catarina and Rio Grande do Sul states (1875–1970)" was initiated, under the coordination of Eunice Sueli Nodari, PhD, with the *Laboratory of Immigration, Migration, and Environmental History* of the Universidade Federal de Santa Catarina, located in Florianopolis, Brazil. The project sets focus on discussing the relation between immigration, migration, and deforestation of the Ombrophilous Mixed Forest.

gested the selection of modern planting techniques, in order to encourage the internal market.

Since that time, to consume more apples in Brazil meant not only to have one more fruit variety in one's diet, but to pay more to be able to eat what Americans ate, to copy recipes published in magazines such as *Reader's Digest*, and abandoning old Brazilian rural food habits to embrace modern ones. This is how Brazil has quickly transformed apples into the number two product in the agricultural import list, following only wheat.[2] In this sense, the planting of apple trees in Southern Brazil was encouraged by the emergence of an internal market for fruit, by a consumption culture based on the urban development, and the institutionalization of agricultural knowledge, besides convergence of public and private interests, embedded with a technical discourse.

The modernization of Brazilian agriculture took off after the Second World War, when a new and strong political idea emerged: Brazilian national developmentalism. As Warren Dean noted, this complex concept embraced all public efforts to shift Brazil from a poor country to an industrialized country. The only way to strengthen the Brazilian internal economy and destroy this "primitive" economic landscape—in the perception of federal, regional, and local authorities—was to substitute foreign goods for domestic ones. This policy of import substitution could solve some Brazilian problems, mainly in agriculture. Some Brazilian dependence regarding agricultural products began to be changed, and Brazilian capitalist farmers emerged as a new group preoccupied with technology increase, modern production, despite its conservative political position in society. This issue has been well discussed in Brazilian historiography since the 1980s, but the main focus rests on the social history of marginalized workers who were excluded from rural modernization. The importance of import substitution in Brazilian agriculture is that it shifted organized social action toward statist modernization.

When we think of Brazilian import substitution policy in agriculture, it is necessary to think of an internal modernization program regarding rural areas. Brazilian governments between the 1940s and the 1960s considered positively the role of American specialists and aid in agricultural modernization, for example through the American International Association for

2 Jó Klanovicz, *Natureza corrigida* (Florianopolis: Universidade Federal de Santa Catarina, 2007).

Economic and Social Development (AIA) and the Point IV Program. AIA offered financial support, training, and technical staff, and the Point IV Program offered modern training to Brazilian people. (It should be noted that Tennessee Valley Authority engineers were also essential to the expansion of Brazilian hydroelectricity in the postwar years.). Building on the work of AIA, between 1948 and 1972, the Brazilian federal government organized the EMBRATER system, a coalition of regional agriculture modernization extension and research organizations that offered technical assistance and training for agriculture modernization inside rural families. After 1972, EMBRAPA (the Brazilian Agricultural Research Corporation) was the major tool for modernization in rural regions. The military government of that time argued that EMBRAPA would create a Brazilian agricultural success story by adopting new technologies that would increase agriculturists' wealth. (Paulus) The institutionalization of modern agriculture in Brazil supported the technological and political conquest of the rural world, creating a new reality where machines were more important than people, and where various sectors worked together to strenghten modern agriculture: the Catholic church, state governments, and private interests, and foreign specialists were considered crucial improving the Brazilian way of modernization. In this sense, it is crucial to think of economic changes in the Brazilian countryside, especially the change from a rural world organized by extraction economy to a rural world organized by a new, strong, and massive capitalist agribusiness. Social and economic elites of that time could observe this eminent transformation in rural landscape.

In the late 1950s, for example, timber businesspeople from mid-western Santa Catarina (covered with OMF)[3], being aware of the diminishing timber forests, discovered orchards as a feasible option. Fraiburgo became the first city in Brazil to have a modern commercial apple orchard with French

3 Detailed studies on the constitution of interpersonal relations between members of private agro-industrial companies with governmental agents, as regards to the agriculture modernization of agriculture in Southern Brazil have been particularly developed in Social and Human Sciences after the 1990s. This shows clearly the importance of such relations for the development of agriculture in the area itself. See Patricia May, *Redes político-empresariais de Santa Catarina (1961–1970)* (Florianopolis: Universidade Federal de Santa Catarina, 1998); Ido Michels, *Crítica ao modelo catarinense de desenvolvimento: do planejamento econômico (1956) aos precatórios (1997)* (Campo Grande: UFMT, 1998); and Eros Mussoi, "Políticas públicas para o rural em Santa Catarina: descontinuidades na continuidade", in *Agricultura e espaço rural em Santa Catarina*, eds. M. Paulilo and Schmidt (Florianopolis: UFSC, 2003).

seedlings, in an area undergoing strong modernization based on financial input, technical support, rural extension, and pesticide use. Within this process, several discourses and practices on 'correcting nature' in the area for the planting of apple orchards emerged and featured prominently in public and private documents from professionals involved in the modernization and further growth of such apple orchards.

From 1962 to 2000, Fraiburgo became the largest apple-producing city in Brazil, which has lifted the former Brazilian necessity to import Argentinian apples. In the 1980s the apple orchards expanded over the remaining natural forests, which brought forth important ecological consequences such as fungi and plagues attacking the roots and leaves in the new orchards. The only way out was to intensify the use of pesticides within this consolidation of the country's orchard cultures. The subsequent, uncontrolled use of pesticides generated concern which was intensely discussed nationwide in July 1989, when a load of apples was apprehended for toxicity analysis and considered contaminated by dicofol (a carcinogetic miticide). Despite the consolidation of apple production in Brazil, this event threatened the national commercialization of apples.

In this context I explore (through technical reports, Brazilian press documents, and oral sources) the nationwide debate on pesticide use on Southern Brazilian apple orchards, which occurred in July 1989 through the Brazilian press, as well as the reactions of apple producers on the accusations, and the artificial conditions created to grow the apple orchards. These issues are lead on to a broader analysis of the ideas of toxicity and 'danger', which by 1989 had started to appear in the press demanding the availability of healthier foods and 'food safety'. In order to discuss this issue, I have used documents from the Brazilian national press which discussed contamination.

Brazilian apples with French/Algerian technology

The idea of producing apples in Fraiburgo had started taking shape in the late 1950s, when sawmill owners in the area formerly known as Butiá Verde noticed that the forest reserves were coming to an end. Under the pressure of their dwindling forests, some businesspeople like René Frey and Arnoldo Frey had begun seeking investment alternatives to remain members of the economic and social elite in the region. In Sao Paulo, the Freys sold

Araucaria pine wooden boxes for Schenk Wineyard, a company which had brought the financially troubled sawmill in liaison with the French-Algerian wine producers Mahler-Evrard, and that had put both parties in direct contact between 1959 and 1962.

Mahler-Evrard and the Freys had converging interests: intended to relocate investment from exploitation to agriculture; the other intended to invest in fruit and wine production in Brazil. The French-Algerian group had the know-how in grape and wine production (although not concerning temperate climate fruit in general), and the capital; the Freys had 5,000 ha of land in Fraiburgo, had an interest in fruit planting, and were aware that the soil and climate of their lands were suitable for producing fruit such as apples and grapes. They created a partnership in which the Freys would invest 1,000 ha in planting temperate fruit and grapes while Mahler-Evrard would supply financial capital for the project. Once in Brazil, the Mahler-Evrard group created the Safra S.A. (*Sociedade Agrícola Fraiburgo S.A.*) in 1962. Safra S. A. planted a 40 ha experimental orchard located 5km away from downtown. Shortly after, Safra S.A. started selling seedlings of apple, pear, peach, plum, and prune, fruits (mainly grapes), and liquors (cognac, wine, and sparkling wine) to all states in Central and Southern Brazil.[4]

The next step in the company was to source capital and technical knowledge. They tried to find some help in France, from the best French planter at that time, Georges Delbard. The pomological meeting on September 1st 1965 in Malicorne/France put the Pepinières Delbard in direct contact with the Evrard family. As Delbard remembered: "they had a pilot orchard in Brazil behaving abnormally, I immediately accepted the invitation they made for me to analyze and expand their orchard. The idea of exploring the fruit planting potentials of the largest country in South America was enticing to me."[5] In his first visit to Safra's experimental orchard in 1966 Delbard concluded that the behavior of apple and pear trees planted there was identical to those planted in Algeria, and that "altitude corrects the effects of latitude".[6] The grape cultivator had provided new varieties to Fraiburgo, suggesting soil and landscape handling techniques and investing capital in expanding Safra S.A. until the 1970s, when he decided to quit the partnership.

This initial expansion of temperate climate orchards in the state, from 1963 to 1973, was marked by converging private (initially until 1968) and

4 Henry Evrard, Interview with *Marlon Brandt* (Fraiburgo, 13 dec. 2003).

5 Georges Delbard, *Jardinier du Monde* (Paris: Hachette, 1986), 569

6 Ibid.

public investments. For example the federal law number 5,106, issued on September 2nd 1966, had authorized individuals and businesses to discount income taxes "up to 50 percent of the tax value due to proved investment in foresting and re-foresting, which may be done by planting forest, fruit, and trees on a large scale within the fiscal year".[7] Apple orchards became the very projects of homogeneous forests covered by this law.

By that time, the problem in producing temperate climate fruit in the area lay in the edaphic and climatic incompatibility between Fraiburgo and the apple varieties being produced in the area. Contrastingly, several companies went to Fraiburgo with large-scale projects, taking advantage of the fiscal incentives for re-forestation that the military government had ensured.

The main public investment instrument regarding temperate climate fruit planting in Fraiburgo in the first stages was a state program for temperate climate fruit planting called Profit (*Programa de Fruticultura de Clima Temperado*), released by the self-sufficient technical support, research, and rural extension organization in Santa Catarina, Acaresc (*Associação de Crédito Rural de Santa Catarina*). Its claims were based on the high expenses in fruit imports and the 'new economic alternatives' for agriculture (or for producers who could afford the public advantages).

The state would then hire Safra S.A. as the exclusive provider for the seedlings that were necessary for the maintenance and expansion of the program. This was, therefore, an appropriate moment to develop temperate climate fruit production, under state support (via Profit, Acaresc, and Empasc) on one hand, and under private investment on the other hand. This process had been marked also by the proselytistic diffusion of the belief in technology as the "controlling and correcting force against nature's flaws", as agricultural technicians from Fraiburgo had claimed. By this stage, it was possible to notice how several environmental limitations had appeared regarding fruit planting projects, as well as the introduction of some new or different relations between human and non-human populations in the apple producing areas.[8] Some enterprises as Reflor Ltda. had begun planting orchards on their own land, as well as on rented, commodity, and condominium lands. Seedlings had been acquired from Safra S.A. In 1969 another Frey family business was created in order to execute fruit planting projects using services from Reflor Ltda. and Safra S.A., called *Renar Agropastorial Ltda.* This company

7 Brazil. Federal Bill #5,106, September 2, 1966.
8 Klanovicz, *Natureza corrigida,* 230.

would plant apple orchards using fiscal incentives, utilizing "resources originated from the mother-business (*René Frey & Irmão Ltda.*) timber exploitation"; in other words, funds from taxes were reapplied in the process of capital accumulation by this family. They all would also use Profit technicians.

But in 1969 Brazil was still an apple-importing country; incentives had been raised in order to achieve future self-sufficiency as regards to the national apple market, based on the Southern production. From the business point of view, the country had at least begun producing the fruit for its internal market, using modern techniques, although exporting was still not a reality. Likewise, any ecological issues could be addressed with technical knowledge, since Fraiburgo was a "vast experimental field", according to Willy Frey.[9]

The apple per capita consumption in Brazil went from 0.65kg/year in 1960 to 1.45kg/year in 1970.[10] Such growth certainly had a relationship with the apple orchard expansion in Fraiburgo. Several Brazilian companies had contacted Reflor Ltda. in order to establish orchards and reforestation projects there, taking advantage of the fiscal incentives offered by the federal government and, of course, intending to pay less taxes. Such companies used areas from Reflor Ltda. itself, in commodity and condominium regimes.[11]

---

9 Willy Frey, "Relatório sobre a situação atual da Reflor," in *Processo de criação do Curso Técnico em Agropecuária da Escola de Segundo Grau Sedes Sapientiae encaminhado para a Secretaria Estadual de Educação*, B. Simonetti (manuscript, Fraiburgo, 1973). Carlos Alberto de Abreu, the administrative manager of Safra S.A. claimed in 1973 that in Brazil the temperate climate fruit such as apples, pears, prunes, and plums went through a rough patch, marked by successive European genetic material imports, as well as adaptations and treatments. Safra S.A. had the national lead in the market, as well as 1,013 ha of temperate fruit trees: Merlot, Cabernet, Trebiano, and Marzenino grapes; St. Rosa and St. Rita plums; and Golden Spur, Red Spur, Golden Delicious, Wellspur, Melrose, Blackjohn, Royal Red, and Willie Sharp apples, among others. In 1973 the commercialization of Safra S.A. reached 414,718 seedlings (395,154 apple seedlings, 12,021 prune seedlings, 4,359 plum seedlings, 794 peach seedlings, and 247 pear seedlings, as well as 1,878 rose seedlings and 265 seedlings of other varied fruit). To conclude his report, Abreu claims that the fruit commercialization in Safra S.A. grew yearly in the same proportion as the yearly consumption. See Carlos A. Abreu, "Histórico da Sociedade Agrícola Fraiburgo," in *Processo de criação do Curso Técnico em Agropecuária da Escola de Segundo Grau Sedes Sapientiae encaminhado para a Secretaria Estadual de Educação*, B. Simonetti (manuscript, Fraiburgo, 1973).

10 Associação Brasileira dos Produtores de Maçã (manuscript 2006).

11 A Reflor Ltda. Report (1973) presents data on 10 reforestation projects using *Pinus taeda*, *Pinus elliottis*, and *Araucaria angustifolia*, undertaken between 1967 and 1970, planting a total of 3,618,750 seedlings in 2,176.34 hectares. By using apple trees as a forest species legally valid for forestation, according to Law #5,106/1996, Reflor Ltda. had assisted 13 fruit culture projects between 1967 and 1973, involving 592.5 ha and 542,200 seedlings.

According to agronomist Jorge Bleicher,[12] fruit cultivation needed more professionals; at the same time, planting 500 ha in only a few years had led to deforestation, roads opened for the machinery, soils drained, as well as large quantities of lime used for soil correction. Considering the fact that during the 1990s, the area planted by the company *Portobello Maçãs S.A.* had used 25 to 35 tons of lime per hectare for soil acidity correction, and that orchards had an average area of 100 ha, we can infer that the 500 ha used by Reflor Ltda. had used approximately 15,000 tons of lime; its residues not only remain in the surface, but also contaminate rivers and aquifers.

"Natural" mistakes

In the initial expansion period (1963–1975), soil acidity 'correction' was not the only problem which had demanded investment and technical intervention on natural environments. The severe reduction of native forests, as well as the ever increasing use of miticides, fungicides, and other agrochemicals, in proportion with the planting area expansion, had reduced the amount of pollinating insects. Every expanded hectare had brought forth natural problems or limitations that had challenged the technical knowledge behind fruit cultivation.

Solving the issue of falling flowers was an important step, which could only be solved based on research by the Israeli specialist Amnon Erez. He concluded that the seedling process occurred at a slower pace in Fraiburgo due to the lack of cold hours (a minimum of 700h per year under 7,2°C).[13] As a solution, he had suggested the intensive use of pollinating bees, as well as the use of chemicals to interrupt plant dormancy. "Now that was wild! A new orchard expansion had begun in 1975, and this time the companies began knocking down the native forests to find space for the new plants. Apple trees began producing 28–30 tons per hectare instead of 2–4 tons per hectare. Profit was guaranteed—based on technology" claims Acaresc agronomist Jorge Bleicher.[14]

In 1973 the company planned to plant 112,800 seedlings in 141 ha for the following year, as well as 240,000 apple trees in 300 ha for 1975.

12 Bleicher, *Interview with Jó Klanovicz,* 2002.

13 Amnon Erez, *Interview with Jó Klanovicz* (Florianópolis, February 12, 2007).

14 Bleicher, *Interview with Jó Klanovicz,* 2002.

Still, one other issue persisted in Fraiburgo: variety adaptation. In the early 1970s, many of the orchards planted mainly *Red delicious* and *Golden delicious*. This issue could only be solved through the research compiled in the book *A cultura da maçã* by Japanese agronomist Kenshi Ushirozawa, who had been part of a technical mission in Santa Catarina between 1971 and 1977. Ushirozawa claimed that producing *Red delicious* and *Golden delicious* in 1,000m average altitude resulted in dry, lumpy fruit, with low commercial quality; at the same time, he suggested substituting these varieties for more precocious varieties such as *Gala* apples.[15] Therefore the *Golden* and *Red delicious* orchards were consistently eradicated as new orchards were planted with two main varieties, *Gala* and *Fuji*.

On the verge of the year 1980, Kenshi Ushirozawa studied carefully every possible aspect of fruit cultivation in Santa Catarina: the shape and behavior of apple trees was presented in schemes, productivity charts, plant distribution graphs, and planting was suggested in areas which could be turned swiftly into a field in Fraiburgo. At the same time, Roger Biau's experimental orchard, paid for by Safra S. A., researched intensively on the adaptation and phytotoxicity of the plants.[16]

In the urge for orchard expansion, and based on the positive results in both public and private pieces of research, Fraiburgo had deforested most of the native forest remaining in Santa Catarina state. Over 1,000 ha of OMF were knocked down every year, between 1980 and 1983.[17] Such devastation had 'isolated' the insects even more, reducing local biodiversity.

The 1970s and 1980s consolidated fruit cultivation in Fraiburgo, based on the rational and mechanized fruit cultivation which had attracted professionals and investment from various sectors, giving the impulse to research. The city population, of little over 2,000 inhabitants in 1967, had gone up to over 15,000 in 1985. Apples were the main economic product. The intensively applied technology could guarantee orchard productivity. Automated

15 Kenshi Ushirozawa, *A cultura da maçã* (Florianópolis: Empasc, 1978).

16 A few examples would be the essays on Gramoxone, Kamex, and Esapon herbicides, which have proved effective in orchards without presenting apparent phytotoxicity and had been freed from narrow-leaved grasses in a second essay on August 18th 1975 (12 years after the first research on the experimental orchard). Cf. R. Biau, *Ofício aos técnicos 1*, Agrícola Fraiburgo S/A, Fraiburgo, August 18, 1975 and *Ofício aos técnicos 2*, Agrícola Fraiburgo S/A, Fraiburgo, August 18, 1975.

17 Carlos E. F. Young, *Is deforestation a solution for economic growth in rural areas? evidence from the Brazilian Mata Atlântica* (Oxford: University of Oxford Centre for Brazilian Studies, working paper CBS-36–2002, 2002).

irrigation methods 'fought against' droughts; hail detection and intervention systems 'fought against' hail, popularly known as 'stone rain', with military strategies (radar systems and cloud bombarding with rockets initially imported from France and Switzerland, and later from the Soviet Union); small fires 'fought against' the frozen dew in September, during the springtime flourishing of plants; rather extreme phytosanitation treatments 'attacked' fungi and diseases. There were also methods applied to change water courses, combat "native insects", as well as very efficient logistics during harvest. All of those aspects enabled the discourse which in 1983 affirmed the success of technicians in "correcting nature's flaws".[18]

Some technicians believe that the process of orchard expansion in Fraiburgo during the 1980s, characterized by the advancing devastation of secondary forests, could jeopardize future production due to the new ecological relationships being established between apple trees and fungi, for example.[19] In fact, not only fungi had begun taking an important part in this story, as opposed to the belief in the technological success in correcting the environment, but also mites, insects, and the weather. The language of technicians and producers had become increasingly belligerent when describing the environment. Expressions repeatedly used, such as "to correct nature's flaws", "powerful machines for environmental correction", "rockets to fight bad weather", to "rationalize the landscape", were terms that had become a part of the technicians' daily life, even in their journals, notes, and planners.

In the 1980s, there was a precarious ecological balance because of problems such as the European red spider mite, fruit flies, apple scab, bitter rot, and bleeding canker. With time, other problems emerged such as the apple leafroller, stained Gala leaves, and brown rot, as well as other re-emerging plagues such as the oriental fruit moth. However, in the history of fruit cultivation in Fraiburgo, relations between humans and apple trees had begun involving bitter rot (*Rosellinia necatrix* (Harting) Berlese), bleeding canker (*Phytophthora cactorum* (Lebert et Cohn) Schroeter), Armillaria root rot (*Armillariella mellea* (Fries) Karsten), crown gall (*Agrobacterium tumefaciens* (E.F. Smith et Townsend)), canker (*Nectria galligena*), apple scab (*Venturia inaequalis* (Cooke) Winter) and the anthracnose (*Glomerella cingulata* (Stoneman) Spaudling and Schrenk). As regards plagues, humans had begun to face the codling moth (*Carpocapsa pomonella* Linnaeus), several types of

18 "Domesticar a natureza", *Veja*, (São Paulo, March 25th 1983), 89.

19 Bleicher, *Interview with Jó Klanovicz*, 2002.

mites, the wooly aphid (*Eriosoma lanigerum* Hausmann), and the San Jose scale (*Quadraspidiotus perniciosus* Comstock).[20]

Brazilian apple contamination in 1989

In 1989, Brazilian apple production provided for the internal market and was being exported to Europe. Therefore, this economic activity had become consolidated in Brazil. The sector companies (*Fischer Fraiburgo Agrícola, Grupo VF, Renar Maçãs, Pomifrai Fruticultura, Portobello Alimentos e Pomigrai Frutas*) were celebrating a harvest of nearly 300,000 tons between February and April that year.

Creating and expanding apple orchards in 1980 Brazil would cost approximately US$10,000 per hectare. The orchard would take three years from planting to its first harvest; however, an average harvest of 28 tons per hectare was enough to cover all costs for its annual maintenance. Therefore, harvest commercialization—bigger and better every year, using investment more often, and more effectively—would cover the investment. Profits from fruit cultivation had caused the local and national associations to invest in public sectors for agronomic research, which had generated a sort of dependence between the state agronomy research institutions and the producing sector. An exemplary case would be the state company for agronomic research of Santa Catarina.[21]

The Brazilian fruit culture scenario in 1989 was as good as it could get, or so it seemed. The apple per capita consumption in Brazil went from 1.9kg/year in 1979 to 2.8kg/year in 1988, and the 31,000 tons harvested in 1979 were small when compared to the 300,000 tons harvested in 1988–1989. The three largest apple-producing cities (Fraiburgo, Vacaria, and Sao Joaquim) held the ten largest temperate climate fruit culture companies in Brazil, employing directly over 20,000 people and many others indirectly.[22]

However, between July and August 1989, a nationwide scandal hit ABPM. Agriculture officials apprehended a load of apples which were supposedly from Guarapuava, Parana state, and from Argentina. An analysis

20 Epagri, *A cultura da macieira* (Florianópolis: Editora da Epagri, 2002).

21 BRDE, *Cadeia produtiva da maçã: produção, armazenagem, comercialização, industrialização e apoio do BDRE na região Sul do Brasil* (Porto Alegre: BRDE, 2005).

22 Ibid.

performed by the state technological institute Tecpar (*Instituto de Tecnologia do Paraná*) came to the conclusion that the load was contaminated with miticide dicofol. This violated the Ministry of Agriculture memorandum from September 2nd 1985 which had prohibited in all national territory the commercialization, use, and distribution of organochlorated agrochemicals due to their potential danger to the environment and to human beings.[23] The news about contaminated apples had a negative impact on the commercialization in Brazil; in the last week of July 1989 fruit producers had begun to report deficits.[24] Dicofol, $C_{14}H_9Cl_5O$, is a result of the hydrolysis of DDT (dichlorodiphenyltrichloroethane, chemically known as 1,1,1-trichloro-2,2-di(4-chlorophenyl)ethane). On July 26th 1989, amidst the polemics on the dicofol-contaminated apples, *Exame Vip* Magazine published the cover story "Poison for dessert", in which the editors tackled the consumption of Brazilian contaminated apples.[25] Until then, a nationwide article would rarely have mentioned any risk of toxicity in fruit; this article in particular claimed that the apple producers were to be blamed for the fact that the Brazilian population was ingesting contaminated fruit. The magazine pointed out that "about 2,500 chemicals are dissolved in a seemingly healthy diet, as well as hundreds of pesticides and dangerous fungi which conspire against good nutrition". However, the story admitted that "it was necessary to ingest a considerable amount of carcinogentic apples in ten years to be in serious risk of developing with a tumor."[26]

The distressing story suggested that the apple contamination was a result of a dosage error in defense agrochemicals, which explained the apprehension of a 300-ton load of dicofol-contaminated apples.[27]

23 Article 1 was set to "Prohibit in all national territory the commercialization, use, and distribution of organochlorated agrochemical products destined to agronomy and other uses: Aldrin, Camphene, Toxaphene, DDT, Kepone, Endrin, Heptachlor, Lindane, Endosulfan, Methoxychlor, Chloramine, Dicofol, and Chlorobenzilate.

24 Tarcisio Poglia, "SC produz 58,47% da maçã nacional", *Diário Catarinense* (Florianópolis, July 30, 1989), 5.

25 "O veneno vai à mesa", *Exame Vip*, (São Paulo, July 26, 1989), 40.

26 Ibid.

27 According to the magazine, the apple producers from Parana and Argentina were responsible for the contamination. In turn, Parana state authorities refused to take the blame, attributing the problem to the acquisition of contaminated fruit originated in Santa Catarina—a common practice amongst fruit producers in Guarapuava, who guaranteed commercialization by bolstering their stock with fruit from other regions.

On the same day, the Gazeta Mercantil[28] newspaper, edited in Sao Paulo and considered to be the main Brazilian business newspaper, printed the headline "Agrochemicals: Santa Catarina has produced contaminated apples". Having received information on the carcinogentic residue on the apple samples analyzed by Tecpar, and being aware that the Parana state authorities had held the producers from Santa Catarina responsible, their story outlined the temperate climate fruit crisis, including some complaints by the former and traditional apple suppliers for Brazil—the Argentinians.[29]

The Brazilian government, under pressure from national producers and fearing the commercialization of contaminated fruit, put an embargo on Argentinian apples in the last week of July 1989. The reaction came immediately: on August 1st that year, Argentina demanded the immediate normalization of the sale of apples to Brazil. It stated that the "Argentinian apple export companies operate rigorously under the sanitation laws established by the Brazilian Ministry of Health, and the authorities claim to be afraid that Brazilian imports may drop, unbalancing the commerce between the two countries".[30]

In a counterpropaganda effort by ABPM, on July 30th the *Diário Catarinense* newspaper (the main press vehicle in the state) had set a page for the issue under the headline "Santa Catarina produces 58.47 percent of the apple production in Brazil". According to journalist Tarcísio Poglia,[31] the growth in Brazilian orchard production and productivity had occurred in the same proportion as the national population had grown, but the rise in apple consumption and production had been jeopardized by the dicofol polemics. The journalist explained that dicofol had been used "in apple trees in several countries such as the United States, West Germany, France, Italy, Sweden, and Argentina. It was used in Brazilian orchards until its prohibition in 1985. However, this agrochemical is still free to use in Brazil for orange and cotton plants".[32] Reproducing words from the ABPM president, agronomist Luiz Borges Junior, he argued that there was a prohibition of dicofol in apple trees, "but the apprehension was a timely case, and the contamination levels

28 "Agrotóxicos: Santa Catarina produziu maçã contaminada," *Gazeta Mercantil*, São Paulo, July 26,1989, 13.

29 "Autoridades argentinas querem normalização das vendas ao Brasil," *Gazeta Mercantil*, São Paulo, August 1, 1989, 9.

30 Ibid.

31 Poglia, *Diário Catarinense*.

32 Ibid.

found were 20 times lower than the limit level defined by the World Health Organization."[33]

In press, the next controversial step would be the way ABPM had countered the accusations of contaminated Brazilian apples, a reaction which was also outlined in nationwide newspapers and magazines. "We must undo this misunderstanding regarding the contaminated apples", said their spokesperson to *Gazeta Mercantil* magazine, in an obvious reference to the article published by the same newspaper on July 26th. The story confirmed that "dicofol had been used in a few plantations in Parana state—but at much lower levels than the accepted limit in any developed country. A part of this amount [of apples] will be lost if the market situation is not reversed until September".

In a nationwide campaign, ABPM denied using the prohibited agrochemical use issues in Brazil. However, agronomy technicians and engineers claim that dicofol had still been used, particularly in Fraiburgo. Agronomy Technician #2 remembers how the product "was widely used. It was dangerous, but good—because it was efficient. We used it all the time. All I know is that many didn't even know it was prohibited, and some companies had a lot of this product in Santa Catarina, Rio Grande do Sul, and Parana. But we knew there was a use limit". Agronomy Technician #3 also claimed to have used dicofol frequently when treating apple trees for mites, in Fraiburgo. On the days the product was applied, "we were very careful, we told the application people all the time to put on all the safety equipment: mask, gloves, overalls. Of course, without overalls and boots they couldn't even drive a tractor, but on those days they had to wear more, they had to put on the mask and gloves".[34]

Agronomy Technician #2 narrates: "I was told to take a bin and a tractor to the chemical deposit and take all the containers that had dicofol. After that, two workers opened a hole with the backhoe loader—away from the orchard and from the rivers, but close to the woods. I remember that I went back and forth many times with tractor and bin for two days to take the containers. After that, the order was to bury it all, and bury it well".[35]

The only possible response for the producers was to reinforce the technical discourse of nature correction, affirming through exaggeration rather than technical language, including the use of literary terms, that the country

33 Poglia, *Diário Catarinense.*

34 Agronomy Technician #3, *Interview with Jó Klanovicz* (Florianópolis, May 12, 2005).

35 Agronomy Technician #2, *Interview with Jó Klanovicz* (Florianópolis, May 12, 2005).

had self-sufficient technology and knowledge to control every process related to apple culture, and that the Brazilian apple came onto the market not only as a product of trees, but also of human, technical labor in a corrected natural world.

It was not a mere coincidence that by the end of 1989, Willy Frey had released the book *Fraiburgo: berço da maçã brasileira* (which means *the birthplace of Brazilian apples*). He started from a basic claim of the apple production in the city, which presented the evolution of a territory described by the author himself in 1973 as "a vast experimental field of temperate climate fruit cultivation" for the "Brazilian capital of apples": "apple trees are quite docile".[36]

Right after describing the tree and its docility, Willy Frey turned his attention to the cold—the most important climate phenomenon influencing temperate climate fruit culture projects in Southern Brazil. Again, the author highlights how technology rules over nature:

> in cold countries, apple trees "sleep" under the snow and awake in the spring, and so they flourish uniformly, making perfect pollination possible. In Brazil, the cold is irregular; the heat makes some of the apple trees wake up before others, due to several factors such as their location in high or low terrain, or due to the varieties planted. This issue was, here, gotten over by using technology.[37]

What else could the author present in this crusade, describing the stage and actions of all the characters that had developed Brazilian fruit culture? Willy Frey appealed constantly to technology in orchard constructions, discussing the technical part of planting and the success of the Brazilian apple production in the 1980s. Of course, the author did not forget to mention deforestation and some ecological consequences of apple planting in Fraiburgo. However, he also attributed a technological solution for the problem, claiming at a later stage that Fraiburgo had no ecology issues—based on another book sponsored by himself and written by Gentila Porto Lopes in that same year. In this the author emphasizes how the city had become an example of ecology conservation due to apple production, because the woods had been expanded as a result of the activity itself [sic].[38]

36 Willy Frey, *Fraiburgo; berço da maçã brasileira* (Curitiba: Vincentina, 1989).

37 Ibid.

38 Gentila P. Lopes, *Glória de Pioneiros: o Vale do Rio do Peixe* (Curitiba: Lítero-Técnica, 1989).

Fraiburgo as a "toxic"-city

It is possible to bring together the claims by Willy Frey in 1989, that orchard expansion had altered the environment and biodiversity in Fraiburgo, and the accusations of pesticide catastrophe which was the main focus of environmentalism in the early 1960s, expressed particularly in Rachel Carson's *Silent Spring* (1962). Perhaps by bringing these two literary works closer, they might serve as a rational exercise into understanding some of the characteristics of the debate on dicofol-contaminated apples in July 1989. The analogy between two ages, two spaces, and two points of view on human and non-human relations allows us to consider the relocation of the issue of agrochemical use from a strictly academic sphere to the public and private spheres, and the coverage of the heated debate on toxicity and the contamination risk of supposedly healthy food—which reflects on a tension, in both meaning and sense, of the successful apple planting in Brazil and its own history.

To a certain extent, the constitution of an environmental way of thinking, marked by this threatened world, takes into account a positive aspect of temperate climate agriculture—which is, in the words of Lawrence Buell, democratizing the debate on agrochemicals in an agriculture characterized in itself by external support, particularly of herbicides and pesticides.[39]

I understand the fact that the publicity given to the dicofol contamination of Brazilian apples uncovered in July 1989 had brought up the issue of human intervention in the natural world in Fraiburgo. This had not only had positive consequences, as regards local and economic development, but there were also negative socio-environmental consequences in the region due to, initially, the great devastation of native forest to plant the orchards, and later due to the increasing amount of "nature-correcting" support in order to obtain profitable apple harvests. Metaphorically speaking, the city had been built as a toxic-city in which planting apples based on the increasingly commonplace application of agrochemical products had become part of the daily life in the field, as well as in the urban area, transforming the city into a toxic space, a territory of agrochemicals which has not yet been studied regarding its actual chemical contamination.

The publicity around the contaminated apples episode, and the reaction of apple businesses to the drop in commercialization of the Brazilian fruit

39 Lawrence Buell, *Writing for an endangered world* (Cambridge: Belknap Press, 2002).

as the sector was consolidating, had strengthened the idea of agrochemicals being dangerous. Particularly after 1960, toxicity became a part of everyday life, yet another of the 'modern anxieties' related to nature. The term 'toxic' has itself acquired a new connotation and shape, carrying a strong emotional load as well as a wide range of meanings. This is due to a few elements inherent in the toxicity discourse. Firstly, the term 'toxic' only has meaning when related to other elements; for example, oxygen is toxic for certain organisms and essential for others. In the same way, products such as salt, chloride, or aspirin in high doses may be toxic for human beings, but are beneficial when ingested in appropriate quantities.

On the other hand, another problem closely related to the 'emotional load' and related to the political issue of agrochemical use, concerns the safety of humans, microorganisms, soils, wildlife, and ecosystems. In this sense, after 1980 in the environment of apple plantations in Fraiburgo humans had begun to be 'bothered' by other artificial agents which had resulted from the very technology used to respond to competition between the apple trees and other non-human characters. That was the case with copper sulfate, lime, benomyl, and captafol, within a "phytosanitation treatment calendar".

In this sense, understanding the landscape is also positioning the environmental imagination in specific contexts as regards to specific times and places. Therefore, the world of modern fruit planting in Fraiburgo would carry in itself the anxieties of modernity, characterized by the crossover between technology and a 'suspicious trust' in progress, which often turned into total disbelief, particularly in critical moments, in the ideas of a productive system or economic activity. In other terms, in a path which is constructed culturally on one side and economically on the other, with landscape and the non-human composing the collectivity which this article refers to, become influential in all spheres.

If diseases and plagues are ecologically linked to apple trees (in an example of non-human relations in environmental history), the historicity of these very occurrences is composed by humans who are invisible, but fundamental in the process of historical constitution of ecological relations as a whole. Expressions such as 'natural tragedies', 'nature's flaws', 'the need to correct what nature has as problems' are reflections of the possible meanings of *landscape* as a term. At the end of the 1980s, the production of apple trees in Brazil presented itself as an efficient, modern, and profitable sector of the economy, developing areas such as Fraiburgo. This process had turned apples into a fetish, forcing the construction of the city based on the connection between the product and political, cultural, and economic destinies inherent to it. However, in the process of constructing the local history on apples, a ghost had systematically hovered above the narratives: in spite of the apple usually being considered a 'salvation' for local economy, it is characterized not for its solidity and superior structure, but as a docile and fragile plant which must be treated, domesticated, and controlled.

Surely the issues existing in Brazilian apple production in the 1980s have been solved—particularly those related to contamination. In a time of re-democratization, after a long period of Brazilian civil-military dictatorship (1964–1985), public opinion had a pivotal part in dealing with all societal issues, including ecological ones like the issue of toxicity and food safety—which caused the state to increase surveillance on fruit-producing companies. On the other hand, the press has also played an important role in encouraging the consumer public to put pressure on the production sector in order to achieve better commercially available products. It is possible to say that, until the 1980s, the Brazilian apples sold on the internal market were of poor quality. However, the improvement of planting techniques as well as the investment in research, storage, commercialization, and apple varieties has increased the fruit standards and aggregated value to the product. The problems of that decade had been solved, and the 1990s would raise other issues amidst an ever-changing agricultural science and an even more demanding consumer market, in both national and international spheres.

Fruit culture in Southern Brazil, based on its eminent rationalization, was the territory of the intense use of support, particularly of agrochemical and potentially toxic products, which brings such agricultural activity into a broad debate on concepts such as food safety, toxicity, and risk in contemporary society, especially if we think of modern plantations and their properties.[40] Being one of the first cultures in Brazil to be intensely rationalized in line with the perspective of a conventional big agriculture aiming for the national and later international markets, the apples represent one of the first profitable temperate climate fruit enterprises in a country which is historically known as tropical.

40 Frank Uekötter, "The Magic of One. Reflections on the Pathologies of Monoculture," *RCC Perspectives* 2 (2011).

Plantations and the Rise of a Society in São Tomé and Príncipe

Marina Padrão Temudo

The social life of São Tomé and Príncipe plantations

The volcanic islands of São Tomé and Príncipe, which gave the name to this African country, are located in the Gulf of Guinea. They were a Portuguese colony for almost five centuries, attaining independence in 1975. With an area of 964 km, it is one of the smallest countries in Africa.[2] It became known during the 19th century for the high quality cocoa produced in plantations under forced-labor conditions. Today, the country still produces cocoa, but no longer has a plantation-based economy; rather, it depends up on external aid and is due to become a petrol-state.[1] The country's society, however, remains forever marked by the legacy of the slave trade and the plantation system that produced a fragmented and stratified social tissue in which some are full citizens and the rest are treated as subjects.

The islands were uninhabited at the time of their discovery by the Portuguese. The harsh equatorial climate was not conducive to the settlement of a large white population, and miscegenation was favored from the beginning.[2] However, not only were the white settlers diverse in their nationality and status, but also the African population was composed of rich African slave traders—who adopted European customs—and by slaves of different ethnic groups and geographical origins.[3]

The marriage of white settlers to African slaves was encouraged by the Portuguese crown through granting 'free' status to women and their Creole

1 J. Frynas, G. Wood and R.S. Oliveira, "Business and politics in São Tomé and Príncipe: from cocoa monoculture to petro-state," *Lusotopie* (2003): 33–58.

2 Francisco Tenreiro, *A ilha de São Tomé* (Lisboa: Junta de Investigações do Ultramar, 1961ª), 60, 68; Isabel Henriques, *São Tomé e Príncipe. A invenção de uma sociedade* (Lisboa: Vega, 2000), 104.

3 Ibid., 12, 36–42, 105–106.

descendants.[4] Early in the sixteenth century, the first white settlers started to be replaced by a generation of indigenous Creoles.[5] This group formed an elite—the "Sons of the land"—that, much later, as a result of their own loss of economic and political importance, would be diluted in another social category, the Forros.[6] The Forros were created by the progressive attribution of letters of manumission, but their numbers increased dramatically with the end of slavery in 1869 and with the abolition of the condition of 'freed' in 1875.

São Tomé and Príncipe was marked by various stages of economic development associated with three major export crops: sugar, coffee and cocoa. Early after settlement, sugar cane was introduced by royal provision, and in the mid-sixteenth century São Tomé amounted to the biggest world producer of this crop.[7] The high labor demands of sugar production contributed to the conversion of the archipelago into a slave trade entrepot. Slaves were either used as labor force for the production of sugar or as a commodity to be exported to the plantations of South America.[8]

Runaway slaves introduced a type of settlement characterized by sui generis forms of social and spatial organization, probably similar to the ones found in their ancestors' villages on the continent. While the sugar plantations and the settlers' villages were located on the lowlands of the north and east of the island of São Tomé, fugitive slaves chose the southern and heavily forested mountainous region to build their villages (*quitembos*) and take refuge (Fig. 1). This group formed by the maroons, called Angolares (and referred to in colonial and post-colonial public space as being "native" to SãoTomé and Príncipe as the Forros), was until the nineteenth century away from contact with other people, and dedicated themselves to activities of gathering and fishing. The constant attacks of these maroons on the plantations contributed to a permanent destabilization, becoming a relevant factor in the migration of the sugar planters to Brazil in mid-seventeenth century.[9] Other determinants of their migration were the poor quality of sugar (due to the difficult storage conditions under high air humidity), the death of

4 Henriques, *São Tomé,* 104; Gerhard Seibert, *Camaradas, clientes e compadres. Colonialismo, socialismo e democratização em São Tomé e Príncipe* (Lisboa: Veja, 2001a), 38.

5 Tony Hodges and Malyn Newitt, *São Tomé and Príncipe. From plantation to microstate* (Boulder: Westview Press, 1988), 19.

6 Tenreiro, *A ilha,* 176–177.

7 Hodges and Newitt, *São Tomé,* 20.

8 Ibid.; Henriques, *São Tomé.*

9 Henriques, *São Tomé,* 110–120.

the sugar cane plantations due to an illness, micro-climate change owing to intensive deforestation, and the constant attacks by the Dutch, English and French.[10] Consequently, from the seventeenth century until its abolition in the Portuguese territories (1836), slave trade remained the most important economic activity of the Creole elite who stayed in the archipelago.[11]

In the words of Tenreiro, the agrarian reorganization that accompanies the introduction of coffee (1787) and cocoa (1822) as export crops gives rise to a 'social upheaval', without any alterations to the structure and organization of land holdings and the labor situation as slavery was replaced for a "servile status".[12] In fact, the abolition of slavery did not lead to the end of the plantation economy as occurred in the Caribbean, though ex-slaves refused to work on plantations due to an association of agricultural work with slavery.[13] Tenreiro[14] argues that the Forros had a "feeling of revulsion for any kind of agricultural work".[15] Consequently, the end of slavery marked the beginning of a long period of import of workers, first from Angola, then also from Mozambique and Cape Verde. New socio-cultural groups were thus created, called respectively the Angolans, the Mozambicans, the Cape Verdeans and the Tongas.[16] Since 1960, Cape Verdeans have constituted the dominant group of laborers in the plantations[17], but unlike Tongas, they have maintained their cultural identity.[18] While Angolans and Mozambicans were forced laborers (whose status was almost similar to that of slaves, and included physical punishments, practical impossibility of return to their countries of origin, and lack of ownership of the children born in the plan-

10 Seibert, *Camaradas*, 39–40.

11 Hodges and Newitt, *São Tomé*, 26; Henriques, *São Tomé*, 81.

12 Tenreiro, *A Ilha*, 230.

13 Seibert, *Camaradas*, 17.

14 Tenreiro, *A ilha*, 231.

15 While the "Sons of the land" constituted the native planters and clergymen, the Forro mainly engaged in arts and crafts.

16 The Tongas consisted of Angolan and Mozambican descendants born in the plantations, who belonged to the plantation owners. Over time, the Tongas will assimilate the Forro culture, although its dilution in this group was very limited and restricted to descendants, recognized as such, of Tonga women and Forro men. (Pablo Eyzaguirre, "Competing Systems of Land Tenure in an African Plantation Society," in *Land and Society in Contemporary Africa,* ed. Richard Downs and Stephan Reyna, [Hannover: University Press of New England, 1988], 340–361, here 346, 347, 351.)

17 Pablo Eyzaguirre, "The Ecology of Swidden Agriculture and Agrarian History in São Tomé," *Cahiers d'Études Africaines* XXVI (1986): 113–129, here 123.

18 Eyzaguirre, Competing, 347.

tations), Cape Verdeans were contract workers escaping from the draught-driven cyclical famines in their own archipelago.

The re-colonization of the islands starting in 1852 was characterized by a process of appropriation of land and political power, which for two centuries had remained in the hands of "Sons of the land". With this process, then, it starts a gradual marginalization and dilution as a social group made up by the native elite.[19] During the initial phase of the settlement in the islands, access to land by settlers was done through royal concessions, in which both dwelling and farming had to be accomplished in up to five years.[20] In the case of freed slaves, access to small plots of land was achieved through their membership in religious brotherhoods.[21]

During the seventeenth and eighteenth centuries, due to the exodus of the sugar plantation owners to Brazil, the land remained in the hands of the "Sons of the land".[22] But with the high concentration of ownership resulting from re-colonization, the property of the natives was mainly reduced to small plantations (*rocinha* de Forro) and small plots (*glebas*) located near towns and villages.[23] Clarence-Smith states that many white planters also lost their properties.[24] He argues that class, more than race, determined the outcome of the struggle for land, and that debt was the main mechanism of ownership transfer. Throughout the history of the country, land ownership was used to define the identity and status of the Forro free person, as opposed to a plantation worker.[25] This symbolic value and use of the land overlapped with any criteria of economic efficiency and determined that the number of owners would remain almost unchanged since 1950, given that the transmissions occurred by inheritance without any updating of land registration.[26] The deployment of plantations in the south began in 1884, and forcibly displaced the maroon Angolares from their lands. They then started

19 Seibert, *Camaradas*, 47, 48.

20 Henriques, *São Tomé*, 77.

21 Eyzaguirre, *Competing*, 344.

22 Tenreiro, *A ilha*, 77.

23 Ibid., 181.

24 William Clarence-Smith, "Creoles and peasants in São Tomé, Príncipe, Fernando Póo and Mount Cameroun, in the nineteenth century," *Reunião Internacional de História de África no terceiro quartel do século xix*, Lisboa, IICT (1989): 489–499, here 493.

25 Eyzaguirre, *The Ecology*, 124.

26 Eyzaguirre, *Competing*, 348–349.

to specialize in fishing (for both consumption and selling) and in sporadic contract work for the clearing of forests in the new plantations.[27]

By 1910, when São Tomé and Príncipe was one of the largest producers of cocoa, 90 percent of the land was owned by large plantation owners and Portuguese banks. Between 1890 and 1921, cocoa production increased in São Tomé and Príncipe, becoming the most important export product.[28] The economy of this small archipelago has thus been dependent, since the beginning of the nineteenth century, on cocoa export under an agrarian structure based on large plantations (*roças*). However, the high costs entailed by the import of contract workers from other colonies, pest attacks, the aging of cocoa plants, and the instability of cocoa prices on the world market weakened the economy of the cocoa plantations, making them less competitive than small family farms of other major producers in Africa.[29] In 1961, Tenreiro spoke of the "fragility" of the islands' economy, stating that "the crisis of the cocoa is the very crisis of São Tomé".[30] A bit later, Sousa compared the agrarian structure of the archipelago with Ghana, Nigeria and Fernando Po, where cocoa production was performed by small family farms, concluding that the productivity of these capitalist enterprises was lower than that of peasants in the neighboring countries.[31] According to this author, a new agrarian structure should be created not only because of economic inefficiency, but also to avoid social tensions.[32] Tenreiro took a similar stance, speaking of an "obsolete agrarian structure", an "unstable society" and of the need "to alleviate the social tension resulting from the marginalization of the majority of local people".[33]

27 Tenreiro, *A ilha*, 133–134.

28 Tenreiro, *A ilha*, 226.

29 Eyzaguirre, *The Ecology*, 122; Eyzaguirre, *Competing*, 345; Pablo Eyzaguirre, "The Independence of São Tomé e Príncipe and Agrarian Reform," *Africana* XXVII (1989): 671–678, 673–674; Pablo Eyzaguirre, "Plantations, State Farms and Smallholders: Cocoa Production in São Tomé," (paper presented at the Cocoa and Development Conference. London, School of Oriental and African Studies, September 15–17, 1999), 3; William Clarence-Smith, "O papel dos custos do trabalho no florescimento e declínio das plantações de cacau em São Tomé e Príncipe," *Revista Internacional de Estudos Africanos* 14–15 (1991): 7–34, 33; Seibert, *Camaradas*, 51.

30 Tenreiro, *A ilha*, 228–235.

31 Alfredo Sousa, "São Tomé e Príncipe. Um caso de concentração," *Estudos Políticos e Sociais* I (1963): 319–336.

32 Ibid., 335.

33 Tenreiro, *A ilha*, 168, 235.

The building of the Santomean society was shaped over time by the organization, hierarchy and economy of the plantations that, in the words of Eyzaguirre, were run as "independent fiefdoms".[34] Fear of no return or of extension of the contracts controlled the conduct of the plantation imported laborers and, as mentioned by Nascimento[35], was one of the factors—alongside with others such as social uprooting and a condition of social inferiority—to determine a behavior shaped by passive resistance, apathy or resignation.[36] Although conferring greater emphasis on individual forms of resistance—such as neglect, desertion, absenteeism, drinking, suicide, theft and sabotage –, Clarence-Smith also acknowledges the occurrence of collective resistance.[37]

The economic crisis of the 1930s led to a gradual effort to force the local population to work on the plantations.[38] Various measures were taken (in particular, between 1945–1953) not only to solve the problem of shortage of labor, but also to improve the infrastructure of the islands.[39] Since 1956 the native underclass has gradually been working in the plantations—including through contracts—located close to their villages, but the revulsion to agricultural work continued to characterize the Forros as a whole.[40] However, they would accept to work in plantations in arts and crafts not directly related to field work. In the mid-1960s, some Forros started to occupy leading positions in the plantations as foremen (capataz) and taskmasters (feitor), and in 1969 about 5,000 Forros were engaged in the plantations work.[41]

After independence, despite the obvious economic difficulties of the plantation system—in a context in which the so-called "natives" refused to collaborate in agricultural work and cocoa prices underwent constant fluctuations in the international market –, the socialist regime of the new state decided to nationalize the plantations and to gather them in fifteen state

34 Eyzaguirre, *The Ecology*,121.

35 Augusto Nascimento, *Mutações sociais e políticas em São Tomé e Príncipe nos séculos xix–xx. Uma síntese interpretativa* (Lisboa: APAD, 2000?), 19.

36 See also Hodges and Newitt, *São Tomé*, 64.

37 Clarence-Smith, *O papel*, 24–25.

38 Nascimento, *Mutações sociais*, 13–14.

39 Seibert, *Camaradas*, 76–80. Forro collective resistance to these measures culminated in 1953 with the so-called Batepá massacre, see Gerhard Seibert, "The February 1953 Massacre in São Tomé: Crack in the Salazarist Image of Multiracial Harmony and Impetus for Nationalist Demands for Independence," *Portuguese Studies Review* X (2002): 52–77.

40 Ibid.

41 Eyzaguirre, *The Ecology*.

enterprises. This decision was justified by the exodus of the former Portuguese managers and technical staff, but also by a need to please a major donor—the Soviet Block. Prior to Independence, however, the anti-colonial party defended a land reform through the distribution of plots to plantation workers and smallholders. According to Eyzaguirre, the concentration of land was throughout history part of a process by which elites and their institutions controlled access to land and to other natural resources as an instrument to ensure their political hegemony.[42] After Independence, the notion of 'autochthony' was used by the Forro elite to justify the appropriation of the state apparatus—and through it the internal and external resources—, while keeping the former plantations' field workers in conditions of political, economic and social subordination. Agricultural workers of state enterprises continued to be the former plantation laborers and their descendants. While most of the Angolans and the Mozambicans did return to their homeland, both the Cape Verdean and São Tomean governments were not interested in the repatriation of the Cape Verdeans. They were forced to stay against their will[43] in the state enterprises, which maintained the key characteristics of the plantations and, according to Eyzaguire, continued "to encapsulate in an inferior status a significant portion of the population".[44]

The nationalization of land and the creation of state enterprises forced the country to remain dependent on the exports of a single crop and on food imports. Notice that in the mid-1980s the country imported 90 percent of its food needs.[45] The persistent fall in cocoa prices since 1979 and the mismanagement to which the state enterprises have been submitted led to its financial collapse and to a heavy external debt of the country.[46] Until 1985, the plots that the plantation workers (mostly the Cape Verdeans) cultivated with food crops since the 1960s to supplement their diet were officially banned.[47] Despite this prohibition, many plantation workers were squatters. Eyzaguirre argues that both for Tongas and for Cape Verdeans, whose social and political links with the Forro elite are weak, "swidden farming represents an opportunity to acquire significant amounts of cash income, which is then

42 Eyzaguirre, *Competing*.

43 Ibid., 677; Seibert, *Camaradas*, 162.

44 Eyzaguirre, *The Independence*, 676.

45 Gerhard Seibert, "São Tomé and Príncipe. Recent History," in *Africa South of the Sahara 2002*, ed. K. Murison (London: Europa Publications, 2001b), 824–840, here 832.

46 Ibid.

47 Seibert, *Camaradas*, 168.

reinvested in a network of kin that may serve to integrate them more fully into the national society".[48] The drought episode of 1982–84 that hit Sub-Saharan Africa affected the country in 1983, creating a huge food shortage. This fact made the need to diversify agricultural production towards increased investment in food crops even more perceptible for plantation workers (but also for the government).

The food crisis triggered by the drought contributed to a change in the Forro attitude towards agricultural work, creating new food habits and also a closer relationship between the Forro of lower status and the former plantation workers belonging to other socio-cultural groups. When, in 1984, the government of Sao Tomé began a process of registration of squatters these farmers selected some Forros to represent them—"owing to the perception that Forros had a favored access to the state and its resources"[49]—and, simultaneously, state plantations' managers having to deal with the squatters selected Forros as leaders, putting them in a position of ascendancy.

In 1985, the São Tomé and Príncipe government began a process of economic liberalization, adopting a structural adjustment program (SAP) in 1987.[50] It was also in the 1980s that the government started a first process (1985–1989) of land distribution—the so-called "citizens' involvement in production"[51] –, as well as a Cocoa Rehabilitation Project associated with the SAP. This first step towards a distributive land reform was conducted in a rather arbitrary way and, in 1989, only about one-third of the area was being cultivated.[52] At the same time, the seven best state plantations were granted with either lease or management contracts, and some of them were subject to heavy funding for the modernization of the technological infrastructure. However, in 1990, the ineffectiveness of the management contracts became apparent, given that the Santomean state reserved the right of supervision over major decisions and, according to the managers, 80 percent of the profits were taken by the state.[53]

The "privatization" of land emerged as a component of the credit agreement with the World Bank, as a result of the failure in the implementation

48 Eyzaguirre, *The Ecology*, 125.

49 Eyzaguirre, *Competing*, 354.

50 Seibert, *São Tomé*, 828.

51 Seibert, *Camaradas*, 202.

52 Ibid., 229; and Seibert, *São Tomé*, 831.

53 Ibid., 830.

of structural adjustment measures.[54] A minimum of 75 percent of the 20,000 ha of land ought to be given to small farmers, and the new land structure should consist of: a) 6,000 smallholders oriented towards food production for self-consumption and to supply the market; b) about 150 medium-size plantations capable to revolutionize and modernize agriculture through the production of new cash crops to export; c) six large companies (those that had previously been the subject of substantial external funding) capable to stimulate cocoa exports.[55]

In late 1991 the government established the criteria and the procedures for land distribution to small farmers and medium-size entrepreneurs. The land would be allocated on usufruct bases for a renewable period of twenty years. Farmers should cultivate it and pay an annual fee, although they were not allowed to divide it by inheritance among their descendants. The size of smallholders' plots should be calculated according to the size of each family and the soil quality.[56] The Bureau of Land Reform (Gabinete de Reforma Fundiária, GRF) was created in order to "check, monitor and evaluate the use of and the improvements made on the land assigned" to each family.[57]

As a consequence of the credit agreement with the World Bank, the Project of Land Privatization and Support to the Development of Smallholders (Projecto de Privatização da Terra e de Apoio ao Desenvolvimento do Pequeno Produtor, PPADPP) began in 1992 with four components: land distribution; reduction of the number of plantation workers; operational and management reorganization of the cocoa sector; and creation of agricultural institutions and support services (FAO and WB 2000). The preliminary assumptions were that land privatization would increase the efficiency of agricultural production, would eliminate the financial losses of state enterprises and would encourage the diversification of agricultural production. However, the PPADPP was not conceived as a project of either land reform or rural development as such, but rather as a project of land distribution to

54 Steve Kyle and Christopher Tanner, *Projecto de privatização agrícola e desenvolvimento de pequenas propriedades. Relatório de Missão* (São Tomé: WB, 1999), 3.

55 Ministério da Economia, *Projecto de privatização agrícola e desenvolvimento de pequenas propriedades. Relatório de Gestão (*São Tomé: Ministério da Economia, 2000), 8–9.

56 World Bank, *Democratic Republic of São Tomé and Príncipe Agricultural Privatization and Smallholder Development Project* (S. Tomé: WB, 1991), 22.

57 Ministério da Economia, *Carta de política agrícola e de desenvolvimento rural* (São Tomé: RDSTP, 1999), 3.

former plantation workers. More than a planned privatization of land, it was a project for the "privatization of agriculture".[58]

The average area of plots to be allocated to small farmers, who in the initial estimates should reach 3.6 hectares, was reduced to 2.5 owing to the increasing demand for land by originally unforeseen beneficiaries, which was reinforced by social and political pressures exerted on the project team.[59] By contrast, the maximum area initially defined for medium-size plantations was increased from 50 to 100 hectares, and could even reach 150 under agreement by the GRF (Decree 73/95). This tendency to escape the credit agreement for the benefit of medium-size entrepreneurs was observed by the donors themselves: "the perceived value of land as a resource consolidated among the ruling elite and the more privileged sections of population, created pressure on the project '.[60] Seibert states that the beneficiaries of the medium-size plantations were politicians, civil servants and businessmen, most of whom had no agricultural experience.[61]

According to the Forro political and economic elite, the land distribution to smallholders corresponds to an extreme fragmentation of the land. Small farmers are perceived as "gatherers" who do not invest in cocoa production, have a subsistence production and are responsible for environmental degradation driven by felling valuable timber on their plots. Empirical research conducted in 2002 and 2005 confirmed that this general opinion is unsubstantiated and that the same critical attitude does not apply in relation to medium-size plantation owners. The latter, however, have received plantations of high productive potential, as well as considerable financial support, without having the management expertise to administer the large areas assigned to them.

The economic viability of small farms was constrained from the outset not only by the minute area of plots, but also by its poor quality (lack of or old cocoa trees, low soil fertility, high slope, forest fallows' parcels), since the most productive areas were given as medium- or large-size plantations. The privatization of land resulted in the simultaneous disengagement of the state in relation to the provision of social services, granted in colonial times

58 FAO and WB, *Projet de privatisation agricole et d'appui au développement du petit paysannat. Contribution à la mission de supervision de la Banque mondiale—Rapport préliminaire* (S. Tomé: RDSTP, 2000), vi.

59 Ibid., 10.

60 Ibid., 6; see also Kyle and Tanner, *Projecto*, 5.

61 Seibert, *São Tomé*, 831.

by the plantation owners and, after Independence, by the Santomean state. In theory, these services should be provided by two agro-business companies (SACs), whose main duties were: a) implementing services such as distribution of food aid, health, education, and training of farmers; b) maintaining and managing all the core technologies of former plantations until its privatization; c) importing and selling agricultural inputs; d) making the purchase, processing and marketing of agricultural products of the new farmers.[62] However, the SACs quickly became undercapitalized and never got to ensure any social services. Support for small farmers—under the National Program of Support for Small Family Farming (PNAPAF) created in 1995—should be provided by various international and local NGOs. Their activities were mainly concentrated on organizing farmers into associations and on providing technical guidance for agriculture and the development of small scale technologies (cocoa, coffee, sun-dried fruit, among others). In 1997, the National Federation of Small Farmers (FENAPA) was established as a NGO to coordinate and represent small farmers, but no efforts were made to enlarge rights to land of former plantation workers and their descendents and to create a more just land structure, nor even to increase the provision of social services in rural areas.

Presently, for those who have no land or whose plot is not big enough to attain family needs, poaching on state-owned land is perceived as a possibility. In contrast, the clearing of abandoned land allocated to medium-size absentee entrepreneurs is viewed as totally illegal, although farmers argue that the government should distribute these abandoned plantations. The lack of land titles does not appear to be a constraint to investment by small farmers. Rather, the fear of failure to pay the annual rent to the state, motivated by the inability to generate sufficient income, is a shadow on the life horizon of these undercapitalized small farmers. More than credit or subsidies, they demand state support for the marketing of their products, the supply of inputs at non-speculative prices and the clearing with chain-saws of their forested plots (the forest fallows that some received and are unable to clear with a machete).

The land structure shaped by colonial and post-colonial plantations not only was linked to a certain regime of exploitation of the labor force that conditioned the productivity of export crops, but also to the making of a fractured society. As highlighted by Eyzaguirre, plantations geared and per-

62 Ministério da Economia, *Projeto*, 18.

petuated an extremely stratified social tissue.[63] In the following section, I discuss how access to land and labor by the plantations also conditioned the farming systems and agricultural experimentation, shaping the landscape and the food regime in ways that left an historical imprint still visible today.

A history of agricultural experimentation: plantations and smallholdings

The Portuguese navigators, when they arrived to the São Tomé island, found dense humid forests of tall trees that looked as if they "were reaching the sky".[64] The two islands are mountainous and volcanic and the climate is tropical humid.[65] Wheat was one of the first crops to be tested—along with vineyards, oranges and other citrus, olive trees, peaches, figs, and many vegetables –, though unsuccessfully.[66] Soon after, the objective of the Portuguese crown became oriented to the transformation of the archipelago into a sugar-producing region, given the favorable climate and soil conditions and the ample availability of water and wood, essential to the operation of the mills. Until then, the production of food crops characterized the island agriculture.

Sugar plantations in the sixteenth century preferred to acquire couples of slaves favoring the adoption of a particular regime of labor force reproduction and access to land. These couples should provide for their own sustenance (food, cloth and shelter), by working on sugar during six days and having one day a week to work on their own fields.[67] Though no historical sources are available and the Portuguese introduced a myriad of exotic species in the islands[68], one may wonder then if the slaves themselves had not also participated in the introduction of new plants, as happened in other

63 Eyzaguirre, *Competing.*

64 Piloto Anónimo cited in Ezequiel Campos, "A Ilha de S.Tomé: Antiga e Actual," *Revista de Estudos Ultramarinos* 5 (1955): 199–231.

65 Tenreiro, *A Ilha.*

66 Ibid., 91.

67 Francisco Tenreiro, "A Floresta e a Ocupação Humana na Ilha de São Tomé," *Garcia de Horta* 9 (1961b): 649–656.

68 José Ferrão, "O ciclo de cacau nas ilha de São Tomé e Príncipe," *Separata de Africana* 25 (2002); Tenreiro, *A Floresta.*

parts of the world.[69] Among the edible plants introduced by the Portuguese until the end of the nineteenth century were bananas, plantain, yams, taro, palm oil tree (*Eleais guineeensis*), *Parinari excelsa*, *Afromomum malagueta*, *Piper guinense*, *Pachylobus edulis*, *Treculia Africana*, *Cola acuminata*, bread fruit, cinnamon, mango, jack fruit, coconut, avocado, pineapple, papaya, sap-sap, guava, sorghum, maize, cassava, peanuts, mustard, beans, tomato and Irish and sweet potatoes.[70] Rice is an interesting case, though the sources do not mention whether the first varieties introduced belonged only to the *O. sativa* species or also to *O. glaberrima*.[71] That is, whether slaves themselves introduced African rice seeds, as it happened in the Americas.

At the height of sugar production, in mid sixteenth century, all forests under 200 meters of altitude had been slashed. This amounted to about one third of the São Tomé island located along the coastline.[72] As Morbey states, the decrease in soil fertility demanded constant clearing of new forests, triggering intense soil erosion and transforming the north of the country into an arid and deforested area.[73] Since the migration to Brasil of the sugar plantation owners until the mid-nineteen century, when coffee and cocoa became the main cash crops, swidden agriculture was the farming system used in São Tomé.[74] Main food crops were taro, yams, sorghum (*milho zaburro*) and bananas, but cotton, ginger and rice were also produced for export in the plantations of the 'Sons of the land'.[75] Tenreiro called this period of almost two centuries "the great fallow", because an intense afforestation process occurred in previous sugar production areas.[76]

At the end of the eighteenth century, the governor of São Tomé and Príncipe introduced the coffee plant[77] with the aim to stimulate the economic revival of the islands, along the lines of what was happening in Brazil.[78] Between 1858 and 1878, skilled coffee planters from Brazil, but also Luso-

69 Judith Carney and Richard Rosomoff, *In the Shadow of Slavery: Africa's Botanical Legacy in the Atlantic World* (Berkeley: University of California Press, 2009).

70 Tenreiro, *A ilha*, 92–93.

71 Ibid., 93.

72 Eyzaguirre, *The Ecology*.

73 Tomás Morbey, "A Actividade Agrária dos Técnicos Portugueses no Território de S.Tomé e Príncipe," *Revista de Ciências Agrárias* 12 (1988): 97–105.

74 Tenreiro, *A ilha*.

75 Hodges and Newitt, *São Tomé*, 24.

76 Tenreiro, *A ilha*, 215.

77 Francisco Rodrigues, *S. Tomé e Príncipe sob o Ponto de Vista Agrícola* (Lisboa: Junta de Investigações Científicas do Ultramar, 1974); Morbey, *A Actividade*.

78 Tenreiro, *A Floresta*.

Brazilian and Santomean ex-slave traders invested in the creation of new plantations and coffee soon became the most import cash crop.[79] Around 1820, cocoa was also introduced as an ornamental plant in the island of Príncipe by a former Brazilian planter and later it became a cash crop. The substitution of coffee by cocoa as the main cash crop was also due to the upsurge of a major fungus attack by coffee rust (*Hemileia vastatrix*). Either coffee or cocoa orchards were, at that time, planted in pure strands with no shadow trees, which later created sustainability problems.[80] Quina tree, hemp (*Canabis sativa*), rubber (*Castilloa elastic*, *Ficus elastica*, and *Hevea brasiliensis*), vanilla, ricin (*Ricinus comunis*), nutmeg, cashew tree, tea, and pepper (*Piper nigrum*) among others, were also introduced during this period.[81]

Between 1885 and 1906 extensive areas of forest had been cleared for the creation/expansion of plantations[82] and some primary forests were also slashed.[83] In 1890, total cocoa exports surpassed those of coffee and, in 1903, the tiny archipelago became the second world cocoa producer after Ecuador.[84] However, since 1920, the combined effect of attacks by cocoa and coffee pests and diseases and the decrease in prices on the international market led to the abandonment of the less productive areas of the plantations. From then on, a period of decline in cocoa production levels due to trips (*Selenothrips rubrocinctus*) attacks started; the intensity of the pest was due to a lack of shadow trees and to the change in the rains regime as a consequence of intensive deforestation.[85] At that time, some plantations' administrators started to allow the laborers to produce subsistence crops on these abandoned lands.[86]

Bananas and palm oil were produced in the plantations intercropped with cocoa and integrated the basic diet of field workers.[87] Though cocoa was the main crop, in general, plantations produced a second important cash crop (mainly copra, palm kernels and palm oil) and several food crops (mainly, bananas, plantain, maize, cassava, beans, taro, bread fruit, jack fruit, mango, palm oil and wine), for the sustenance of the white personnel, field labors

79 Campos, *A Ilha*.

80 Rodrigues, *S. Tomé*.

81 Ibid., 63–64.

82 Campos, *A Ilha*.

83 Tenreiro, *A Floresta*, 84.

84 Campos, *A Ilha*.

85 Ibid.

86 Clarence-Smith, *Creoles*.

87 Tenreiro, *A Floresta*.

and livestock. Some food crops (bread fruit, jack fruit, cola, mango, bananas, plantains, palm oil trees, among others) were intercropped with cocoa and coffee as part of a multi-layered shadow agro-forestry system.[88] Cattle, sheep, goats, chickens, ducks and even rabbits were also raised. However, lack of labor force was a major constraint and the plantations were highly dependent on the import of foodstuff. To face this problem, plantation managers became more tolerant in relation to granting marginal lands to field workers. Nonetheless, only loyal workers would be entitled to plots, as access to land was used as a weapon to enforce compliance.[89] In these plots, workers planted food crops complementing the plantation diet with more desirable food.

After World War II, the price of raw materials increased, particularly that of oils, inducing agricultural diversification in the plantations.[90] The distribution of the cash crops along the landscape was dependent upon the agro-ecological conditions related to altitude[91]: coconut and palm oil trees were cultivated along the coastline, then cocoa and coffee (the latter up to 1,000 meters altitude), either *C. arabica* (in the north) or *C. liberica* (in the south). Above this altitude there were still some plantations of quina trees' (*Cinchona officinalis*). In the small plantations and *glebas* of the Forros, mainly located in the lowlands of the north of the country around the capital, polyculture in agroforestry was also practiced in which cocoa (and sometimes coffee) was produced for export under the shadow of palm oil trees, bread fruit, mangoes, bananas, plantain and jack trees, among others. Other food crops such as yams, cassava, sweet potatoes and several spices were cultivated and small livestock was also raised. Cocoa, coffee and palm oil were processed by the Forros with rudimentary technological methods.

The new fall in cocoa prices since 1955 forced the abandonment of the more distant or less productive lands of the plantations, which since then entered in a long fallow period.[92] Sousa studied the agrarian structure of the archipelago in relation to land productivity in the plantations of more than 4 ha.[93] The author concluded that: only around 25 percent of the total area is being properly farmed; 63 percent of the plantations (with 10–500 ha) cover an area of only 7 percent and, on the contrary, the 23 biggest planta-

88 Rodrigues, *S. Tomé*, 69, 76.
89 Eyzaguirre, *The Ecology*.
90 Tenreiro, *A Floresta*, 154.
91 Tenreiro, *A Ilha*.
92 Tenreiro, *A Floresta*.
93 Sousa, *São Tomé*.

tions (with more than 1000 ha) held 85 percent of the total area. As previously mentioned, Eyzaguirre argues that the small plantations and *glebas* of the Forros had a symbolic value associated with the status of non-plantation workers, which surpassed the importance of the productive use of the land.[94] As such, the land was not divided between heirs and a variety of people claimed ownership and use rights. Usually that meant that land productivity was not an aim and that the plots were worked by many people who could then have occasional access to the agricultural products.

After Independence, plantations were nationalized and the former manager and technical staff were replaced by Forros with no previous expertise in tropical agriculture and management. The work organization characteristic of colonial plantations also eroded. Cocoa trees ceased being replanted, pruned, fertilized and watered regularly and their productivity declined dramatically. New areas were also abandoned. The meals provided to workers became even worse and many of them started to open plots in the fallows in order to secure their sustenance.[95] Maize, beans, sweet potatoes, cassava (traditional ingredients in the Cape Verdean diet), tomatoes, taro, cabbages and plantains were produced in these fields (*campo* or *lavra*). Full-sun cocoa hybrids from Ivory Coast were introduced, but they showed poor results in terms of productivity. A minute use of agrochemicals may explain the failure in the adoption of these hybrids, in stark contrast with what happened in the neighboring African countries.[96]

Eyzaguirre studied the gradual emergence of squatters of different cultural groups on forest fallows (*capoeira*) owned by the plantations and on lowland marshes near the city.[97] While shifting cultivation was practiced in the forest fallows, intensive agriculture was carried out in the lowlands. Comparing the achievement of the plantations' farming systems with that of the squatters, the author concludes that squatters are the ones to develop a market-oriented agriculture, whose decisions regarding the distribution and allocation of resources are based on assumptions of market dynamics and production costs.[98] However, the need to adopt a low profile so as to avoid the greed of the state becomes a factor limiting the expansion of acreage and

94 Eyzaguirre, *Competing*, 346–349.

95 Hodges and Newitt, *São Tomé*, 66.

96 François Ruf, "The myth of complex agroforests: the case of Ghana," *Human Ecology* 39 (2011): 373–388.

97 Eyzaguirre, *The Ecology*; and Eyzaguirre, *Competing*.

98 Eyzaguirre, *Competing*, 355.

productive investment by the squatters. Some cases have occurred in which the state has expropriated very productive plots, just to leave them again abandoned.[99]

During land distribution under the Project of Land Privatization and Support to the Development of Smallholders (PPADPP), plots were not attributed according to planned criteria, and the best ones were granted to privileged members of the plantation administrative staff.[100] Some former plantation field workers did not receive land and were integrated into large and medium-size plantations, while others only received rather small plots (1 ha), and many received old forest fallows, with steep gradient and far from the house (more than one hour walk on a hilly and slippery trail). With the exception of the high altitude areas, both the landscape and the farming systems continue to be, as during the plantations era, relatively homogeneous and constituted by few crops produced in agroforestry.

Revealing the failure of the land privatization program to achieve cash crop diversification, Seibert states that cocoa still represented, in 2004, about 91 percent of exports and 60 percent of the arable land.[101] This, however, conceals a more complex reality. On the one hand, most of the medium-size plantation owners, who allegedly were meant to diversify production into new cash crops for export, looted their farms of whatever was left since Independence (timber, cocoa, palm trees) and did not make use of the credit and training received to invest in agriculture. On the other hand, although smallholders had been given cocoa plants and shadow trees to plant after land privatization, cocoa is rather capital intensive in labor and inputs (diverse agrochemicals). The smalholders lacked capital, and also the knowledge and interest in doing the technological transformation of cocoa (due to high labor demands and lack of financial return), and started to treat it as a wild plant whose fruits are collected when it is profitable. Furthermore, the six big well-endowed plantations—of the pyramid-like land structure imposed by the World Bank in 1992—which were meant to guarantee the country's revenues from cocoa export did not resist poor management and citizens'

99 Eyzaguirre, *Competing*, 353.
100 Helmle cited in Seibert, *Camaradas*, 330.
101 Gerhard Seibert, "São Tomé and Príncipe. Recent History," in *Africa South of the Sahara 2007*, ed. Ian Frame (London: Routledge, 2006), 972–987.

pressure upon the government to distribute land and started to be divided into medium-size plantations and smallholdings.[102]

Early sugar cane plantations left an imprint in the agro-ecological conditions of the country, whose north presents a savannah land cover. Still, in 2001, as a component of PNAPAF, the program for the development of an "Organic Cocoa Commodity Chain" was started in the northern, drier region of the country, so as to avoid problems with brown rot. In 2003 the program continued under PAPAFPA, which funded the construction of infrastructures for cocoa technology and the purchase of some tools, so that the farmers could control the whole production-transformation chain. However, this alleged fair trade commodity chain is not perceived as profitable enough by farmers, who engage in it only because of their lack of alternatives for a regular cash income source. In fact, if the price is relatively superior to that of unprocessed cocoa beans, the production constraints imposed by organic agriculture reduce drastically the output and in parallel highly increase labor demands. Furthermore, farmers are prevented from selling rough cocoa beans (if caught they will be expelled from the program) when they are in need of an immediate cash income for an unscheduled expenditure, such as treatment for an illness.

The sharp binary opposition between Forro and former plantation field workers in their relative perception of agricultural work is becoming blurred in the case of poor Forro, who now see land more as a source of sustenance than of cultural distinction. But while there is a growing number of lower class Forro who maintain a relationship with the land which is very similar to that of most Cape Verdeans, there is also a different attitude, perceived as lazy and careless—and regarded as characteristic of the Forros—among many young Cape Verdeans of second and third generations. Four main production strategies were found among the interviewees: Firstly, the group formed by the majority of interviewees, which includes farmers of all social categories (from Forro to second and third generation Cape Verdeans). They make a small labor investment on the plot by planting taro, plantain and bananas, and collecting cocoa when the price pays off and buyers come to the community. They may also engage in other short term cash-income activities, such as palm wine tapping and charcoal production. Secondly, the

102 José Mendes, *Avaliação do 1ºciclo do PAPAFPA* (São Tomé: Ministério da Economia, 2005); Ministério da Economia and FAO, *Actualização da carta de política agrícola e do desenvolvimento rural. Documento de trabalho GT 13: o processo fundiário* (S.Tomé: RDSTP, 2006), 3–4.

group formed by those embraced by the organic cocoa project (in São Tomé island) and those able to access a regular buyer for the rough beans at a relatively fair price and who have few market opportunities for other productions (in Príncipe island). Although their main cash income source may be cocoa, they also engage in the production of other crops for home consumption and for the market whenever they have connections that allow them to send surplus to the São Tomean market. Thirdly, the absentee or semi-absentee farmer. They belong to the Forro elite's clientele, have a job and live in a village or the capital city. The land is usually rented for agriculture or for palm wine tapping and timber is being slashed for sale. Some of them hire local youths to cut timber in the forest by a chain-saw, which they then sell in the city. Semi-absentee farmers come on weekends, most have permanent or casual workers and mainly produce taro and bananas. Fourthly, the last category is formed by some highly innovative farmers, who distinguish themselves from the others by their entrepreneurial attitude, hard work, agricultural knowledge and experimental skills and diversified livelihoods. They are frequently testing new plants, new techniques, and/or new niche markets. At present, some of these farmers are becoming comparatively wealthier, and have started to buy or rent plots from farmers of the other groups and to hire permanent or casual workers, so as to be able to expand the production area and/or invest in more intensive production techniques or crops. For all of these small farmers, independently of the category to which they belong, the diet is constituted by taro, bananas/plantain and bread fruit and the main source of proteins is a new pest (a huge snail) allegedly imported from the African continent.

In sum, if land tenure and use and, at the same time, a revulsion for agricultural work were self-maintained as enduring characteristics of the Forros—raising them above plantation workers—after Independence land became progressively also a source of livelihood for the poor strata. In their small plots, both the Forros and the non-Forros cultivate taro and bananas/plantain as basic staples and the vast majority uses the old cocoa trees to collect the seeds when the prices are good and buyers come to rural areas. The secondary forests that many decades ago had been cultivated in plantations still produce a myriad of crops whose fruits can be gathered to complement the diet and even the cash revenues of the households.

Conclusion: Plantations and the value of land

Upon its colonization by the Portuguese, about 500 years ago, the islands of São Tomé and Príncipe became known for their plantation-based economy and land structure, in which first sugar cane, then coffee and finally cocoa were the dominant cash crops. Maroon "communities" (*Angolares*) and then the freed slaves (*Forros*) lived at the margins of what constituted a sort of independent 'plantation chiefdoms', adopting their own livelihoods.

Plantations were fashioned as disciplined spaces controlling people and nature. Whereas slavery and slavery-like conditions characterized the history of colonial plantations, the landscape was also entirely transformed by the introduction of new plants and techniques. Late colonial plantations could also be seen as "high modernist schemes"[103]—though ruled with medieval regimes of power –, in relation to architecture and the use of "scientific" knowledge, while outside them the Forro and the Angolares took profit of what a generous nature could offer, without trying too much to transform it in their very small plots. Extreme land concentration in the plantations and the legacy of slavery regarding access to land of the so-called "natives" resulted in a symbolic value of land associated to social status. Plantations' administrators and owners also nurtured a kind of "native-stranger" opposition between the Forro and the plantations' workers to better impose their rule upon them.

If the heyday of colonial plantations was achieved at the cost of imported contract workers from Angola, Mozambique and Cape Verde, the process of socialization of the economy that succeeded independence served the "native" Forros to take over the administrative apparatus of state enterprises and then extend it. With staff having either any managerial or technical skills, with no financial capital to use chemical inputs and fuel, with no authority to enforce disciplined working regimes—when salaries and plantation social services were decreasing –, and facing increasing costs due to an expanded administrative apparatus, state plantations gave the final blow to an institution that had proved inefficient and unprofitable since many decades earlier.

Although São Tomé and Príncipe islands were not inhabited at the time the Portuguese arrived, the politics around putative autochthony were a hidden element across the history of the archipelago's land structure. The land privatization process, implemented in the 1990s with the creation of small-

103 James Scott, *Seeing like a State* (New Haven: Yale University Press, 1998).

holders' "communities", fashioned new inequalities in land tenure. It is the participation in the administrative work of the plantations that will legitimize Forros' access to plots of land during land reform. Further, the Forros closer to the political elite were granted the so-called medium-size enterprises, located in the youngest and most productive cocoa areas. The economic viability of the family farms of ex-plantation labors—those possessing agricultural knowledge and experience—becomes compromised. Having received small plots of low productive potential and facing an unstable cocoa market, only the farmers embraced by the organic cocoa project (north of São Tomé island) and those located near cocoa buyers (mainly in Príncipe island) and unable to produce or commercialize other cash crops were willing to give continuity to the typical São Tomean plantation farming system based on cocoa cultivation under shadow trees. The majority of small farmers, however, feel defeated by the harsh conditions to produce and sell this former dominant cash crop, and treat cocoa as a wild plant whose fruits are collected when prices are favorable. In sum, the new land structure fashioned by the land reform did not solve the problem of the economic viability of the former plantations' main cash crop, and resulted in a big challenge in terms of former plantation workers' livelihoods, but mostly in the access to land of future generations for whom migration seems the only chance of survival. In sum, the enduring stigmatization of former plantation workers remains a salient characteristic of the archipelago's politics, legitimizing their condition of political, social and economic subordination.

Acknowledgements

Field research in São Tomé and Príncipe was funded by the Portuguese Fundação para a Ciência e Tecnologia (FCT). Acknowledgements also go to the Department of Anthropology, University of Yale, USA, where I was a Visiting Fellow when this manuscript was written. To my colleagues and friends António Vieira, Jaime Duarte, Gerhard Seibert, Alexandra Arvéola, Arlindo Lima, Severino Espírito Santo, António Salgueiro, Cristiano Dôndo and Tomé Santos for their support and share of ideas and documents. The poor smallholders of São Tomé and Príncipe who replied to my endless questions must however receive my deepest acknowledgements.

Notes on Authors

Michitake Aso is an assistant professor at the Department of History of the University at Albany–State University of New York, USA.

Christiane Berth is a research assistant at the Institute for European Global Studies at the University of Basel in Switzerland.

Jó Klanovicz is an assistant professor at Midwestern State University of Parana, Brazil.

Stuart McCook is an associate professor at the Department of History of the University of Guelph in Ontario, Canada.

Andrew McWilliam is an associate professor at the College of Asia and the Pacific of Australian National University in Canberra, Australia.

Michael Roche is a professor at the School of People Environment and Planning of Massey University in Palmerston North, New Zealand.

Chris Shepherd is a postdoctoral fellow at the School of Culture, History and Language of Australian National University in Canberra, Australia.

Mart A. Stewart is a professor in the Department of History at Western Washington University in Bellingham, USA.

Marina Padrão Temudo is a researcher at the Tropical Research Institute in Lisbon, Portugal, and the African Studies Centre at the University of Oxford, United Kingdom.

Frank Uekötter is a Reader in Environmental Humanities at the University of Birmingham, United Kingdom.

Jeannie M. Whayne is a professor at the Department of History of the University of Arkansas in Fayetteville, USA.

Index

Social Science

Stefan B. Kirmse
Youth and Globalization in Central Asia
Everyday Life between Religion, Media, and International Donors
2013. 337 pages. ISBN 978-3-593-39889-1

Marina Hennig, Ulrik Brandes, Jürgen Pfeffer, Ines Mergel
Studying Social Networks
A Guide to Empirical Research
2012. 218 pages. ISBN 978-3-593-39763-4

Matthias Bergmann, Thomas Jahn, Tobias Knobloch, Wolfgang Krohn, Christian Pohl, Engelbert Schramm
Methods for Transdisciplinary Research
A Primer for Practice
2012. 294 pages. ISBN 978-3-593-39647-7

Franz Höllinger, Markus Hadler (eds.)
Crossing Borders, Shifting Boundaries
National and Transnational Identities in Europe and Beyond
2012. 355 pages. ISBN 978-3-593-39612-5

Walter R. Heinz, Johannes Huinink, Ansgar Weymann (eds.)
The Life Course Reader
Individuals and Societies Across Time
2009. 591 pages. ISBN 978-3-593-38805-2

Debra Hopkins, Jochen Kleres, Helena Flam, Helmut Kuzmics (eds.)
Theorizing Emotions
Sociological Explorations and Applications
2009. 343 pages. ISBN 978-3-593-38972-1

Frankfurt. New York